Friction-Induced Vibration in Lead Screw Drives

Orang Vahid-Araghi • Farid Golnaraghi

Friction-Induced Vibration in Lead Screw Drives

Springer

Orang Vahid-Araghi Ph.D
Queen Street N. 175
N2H 2H9 Kitchener Ontario
Apt. 704
Canada
ovahidar@engmail.uwaterloo.ca

Professor Farid Golnaraghi
Esquimalt Ave. 912
V7T 1J8 West Vancouver British
Columbia
Canada
mfgolnar@sfu.ca

ISBN 978-1-4419-1751-5 e-ISBN 978-1-4419-1752-2
DOI 10.1007/978-1-4419-1752-2
Springer New York Dordrecht Heidelberg London

Library of Congress Control Number: 2010938368

Printed on acid-free paper

Springer is part of Springer Science+Business Media (www.springer.com)

Preface

Lead screw drives are used in various motion delivery systems ranging from manufacturing to high precision medical devices. Lead screws come in many different shapes and sizes; they may be big enough to move a 140 tons theatre stage or small enough to be used in a 10-ml liquid dispensing micro-pump. Disproportionate to the popularity of lead screws and their wide range of applications, very little attention has been paid to their dynamical behavior. Only a few works can be found in the literature that touch on the subject of lead screw dynamics and the instabilities caused by friction. This monograph aims to fill this gap by presenting a comprehensive study of lead screw dynamics focusing on the friction-induced instability in such systems.

This book is based partly on the first author's Ph.D. research at the University of Waterloo, Ontario which was carried out under the supervision of the second author. The need for a dedicated and detailed study on the friction-induced vibration in lead screws became evident to the authors when they encountered two lead screw noise problems – over a short period of time – from two very different commercial applications which shared many resemblances. One of these two cases is discussed in Chap. 9.

After a brief introduction to the topic of friction in machines and mechanisms in Chap. 1, some basic information regarding lead screws are presented in Chap. 2. In this chapter, the kinematic relationship between lead screw and nut and the contact forces are introduced which serve as the basis for the mathematical models of Chap. 5. Some mathematical background topics are reviewed in Chap. 3. Included in this chapter is a brief introduction to the mathematical tools used throughout this book; namely, the eigenvalue analysis method and the method of averaging.

Friction can cause instability in dynamical systems through three distinct mechanisms: (1) negative damping, (2) kinematic constraint, and (3) mode coupling. Chapter 4 is dedicated to the introduction of these mechanisms. Illustrative examples are worked out in this chapter to demonstrate the techniques that are applied to the lead screw drives in the later chapters.

A number of mathematical models are developed for lead screw drive systems in Chap. 5. Starting from the basic kinematic model of lead screw and nut, dynamic models are developed with varying number of degrees of freedom corresponding to the different components of a real lead screw drive from the rotary driver (motor) to the translating payload. In these models, dry friction between meshing lead screw and nut threads constitute the sole source of nonlinearity.

Chapters 6–8 are the three main thrusts of this monograph. Negative damping instability mechanism is treated in Chap. 6. Using a 1-DOF dynamic model of a lead screw drive, the destabilizing effect of decreasing coefficient of friction with relative sliding velocity between meshing threads is discussed in detail. The method of first-order averaging is used in this chapter to expand the results of the linear eigenvalue analysis and to explore the existence and stability of periodic solutions. In Chap. 7, the mode coupling instability mechanism in lead screw drives is considered. A number of multi-DOF models – developed in Chap. 5 – are used in this chapter to explore the conditions under which vibrations (due to mode coupling) can occur in a lead screw drive system. The kinematic constraint instability is the subject of Chap. 8. Based on the results of Chap. 4, the connection between the well-known Painlevé paradoxes and instability is highlighted. In this chapter, parametric conditions for the onset of kinematic constraint instability are derived.

A practical case study is presented in Chap. 9 where friction-induced vibration in a lead screw drive is the cause of excessive audible noise. Using a complete dynamical model of this drive, a two-stage system parameter identification and fine-tuning method is developed to estimate the parameters of the velocity-dependent coefficient of friction model. The verified mathematical model is then used to study the role of various system parameters on the stability of the system and on the amplitude of vibrations. These studies lead to possible design modifications that can solve the system's excessive noise problem.

The current work provides a detailed treatment of the dynamics of lead screw drives and the friction-induced vibration in such systems. The reported findings regarding the three instability mechanisms and the friction parameters identification approach can improve the design process of lead screw drives.

Waterloo, Ontario — Orang Vahid-Araghi
Vancouver, British Columbia — Farid Golnaraghi

Contents

Chapter 1
Introduction

Lead screws are used in various motion delivery systems where power is transmitted by converting rotary to linear motion. Packaging industries, industrial automation, medical devices, and automotive applications are some of the areas where lead screws can be found. Lead screws come in many different shapes and sizes; they may be large enough to support and move a 140-ton theatre stage [1], lightweight enough to be considered for wearable robotic applications [2], or even small enough to derive micropumps used in medical applications to dispense fluid volumes of less than 1 ml with precision [3].

The sliding nature of the contact in lead screw drives puts great importance on the role of friction on their performance. In addition to efficiency concerns, driving torque requirements, or wear, friction can be the cause of dynamic instabilities, resulting in self-excited vibrations which deteriorates the performance of the system and may cause unacceptable levels of audible noise.

Numerous researchers have studied self-excited vibration phenomena in a variety of frictional mechanisms (see, e.g., [4–9]). A considerable portion of the research in the field of friction-induced vibration is devoted to the brake systems. See, for example, the review paper by Kinkaid et al. [7]. The major self-excited vibration mechanisms in the systems with friction can be categorized into three types:

1. Decreasing friction force with increasing relative velocity or negative damping
2. Mode coupling
3. Kinematic constraint

The purpose of this monograph is to present a comprehensive study of the dynamics of the lead screw drives focusing on the effects of dry friction. Using a unified framework for the dynamic modeling of lead screw drives, the above three friction-instability mechanisms are studied in detail. For each instability mechanism, instability conditions are derived and the vibratory behavior of the system is studied.

The results presented throughout this book will help designers better understand the intricacies of the lead screw dynamics with friction and will provide guidelines to prevent friction-induced self-excited vibrations at the design stages as well as in the practical situations.

O. Vahid-Araghi and F. Golnaraghi, *Friction-Induced Vibration in Lead Screw Drives*,
DOI 10.1007/978-1-4419-1752-2_1, © Springer Science+Business Media, LLC 2011

In a case study (Chap. 9), an actual product is studied where friction-induced vibration in the lead screw drives resulted in unacceptable levels of audible noise. An important part of this study is the development of a novel approach to identify system parameters including the velocity-dependent coefficient of friction between lead screw and nut.

1.1 Friction in Machines and Mechanisms

A historical review of structural and mechanical systems with friction is given by Feeny et al. [10]. Their paper starts from the first human experiences in fire making and early inventions of the ancient cultures to the works of *Leonardo da Vinci*, and expands to the modern-day scientific advances in friction utilization and prevention.

An essential part of any study on the behavior of a dynamical system with friction is to appropriately account for the friction effects using a sufficiently accurate friction model. There are numerous works found in the literature on the various friction models for simulation and analysis of dynamical systems. In one of the first survey papers on friction modeling by Armstrong-Helouvry et al. [11], various friction models are studied. These models can be divided into the following two categories:

- Models that are based on the micromechanical interaction between rough surfaces and aim to explain the friction force.
- Models that incorporate various time or system-dependent parameters to reproduce the effect of friction in a dynamical system.

The latter category is the subject of numerous works as can be seen in review papers by Ibrahim [4], Awrejcewicz and Olejnik [5], and Berger [12]. As reported in these works, friction can be considered dependent on any of the following factors: relative sliding velocity, acceleration, friction memory, pre-slip displacement, normal load, dwell time, temperature, etc.

The friction models used in the dynamic modeling of systems can be further divided into *static* models and *dynamic* models. In the dynamic friction models such as the so-called LuGre model [13], the friction force is dependent on additional state variables that are governed by nonlinear differential equations stemming from the model for the average deflection of the contacting surfaces. At the price of increased complexity of the overall system dynamics, these models are capable of reproducing various features of friction such as velocity and acceleration dependence, pre-slip displacement, and hysteresis effect.

Depending on the specific problem being investigated, an appropriate friction model should be chosen that reflects the relevant features of the physical system. The simplest approximate friction model may be given by (see, e.g., [11, 12])

$$F_{\mathrm{f}} = \mu(v) N \operatorname{sgn}(v),$$

Fig. 1.1 Stribeck curve [11]

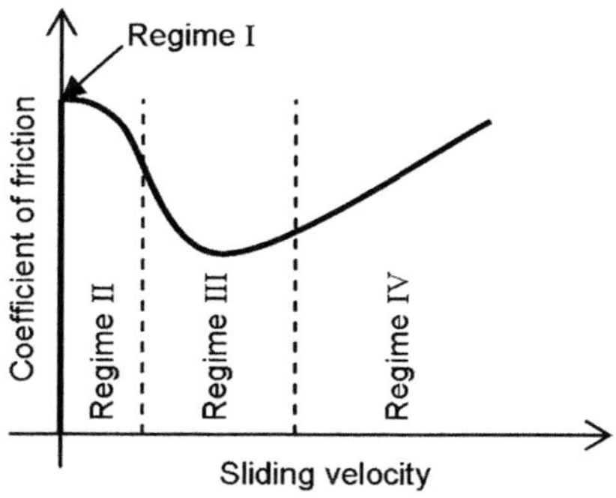

where F_f is the friction force, v is the relative sliding velocity, $\mu(v)$ is the velocity-dependent coefficient of friction, and N is the normal force pressing the two sliding surfaces together. This model is extensively used in the study of friction-induced vibration.

When some form of lubrication is present between the sliding bodies, the variations of friction with velocity is typically explained by the Stribeck curve [14]. As shown in Fig. 1.1, four different regimes are identified in this model [15]. The first regime is the static friction where lubricant does not prevent the contact of the asperities of the two surfaces and friction acts similar to the no lubricant situation. In the second regime, the sliding velocity is not enough to build a fluid film between the surfaces and lubrication has insignificant effect. In the third regime with the increase of velocity, lubricant enters the load-bearing region, which results in partial lubrication. In this regime, increasing the sliding velocity decreases friction. Finally, in the fourth regime, the solid-to-solid contact is eliminated and the load is fully supported by the fluid. In this regime, the friction force is the result of the shear resistance in the fluid and increases linearly with velocity.

Different models have been proposed to reproduce this type of velocity-dependent friction (see, e.g., [16, 17]). The important feature of these models is the existence of a region of negative slope in the friction-velocity curve, which may lead to self-excited vibrations. This type of instability is discussed in Sect. 4.1 and Chap. 6.

1.2 Friction in Lead Screw Drives

Olofsson and Ekerfors [18] investigated the friction-induced noise of screw-nut mechanisms. They discussed the tribological aspects of lubricated interaction between lead screw and nut threads, which accounts for the Stribeck friction. Based on experimental results, they have concluded that: (a) the squeaking noise is the result of self-excited vibration between lead screw and nut threads; (b) in the system studied (consisting of a long and slender screw), these vibrations excite bending mode shapes of the lead screw; and (c) the squeak noise is generated only when the nut is in the vicinity of one of the nodes of the bending mode shape of the lead screw.

In a study of the effect of friction on the existence and uniqueness of the solutions of the equation of motion of dynamical systems, Dupont [19] considered a 1-DOF model of a lead screw system. He investigated the situations under which no solution existed and clearly identified one of the sources of instability in the lead screw systems; i.e., the kinematic constraint instability mechanism. For the self-locking screws, he found that there is a certain limiting ratio between the lead screw moment of inertia (rotating part) and the mass of the translating part, beyond which no solution exists.

Based on a case study, Gallina and Giovagnoni [1] discussed the design of screw jack mechanisms to avoid self-excited vibration. They developed a 2-DOF model of a lead screw system which included lead screw rotation (coupled with the nut translation) and lead screw axial displacement. Using eigenvalue analysis of the linearized equations, they found relationships that define the stability domain in terms of the parameters of the system. They concluded that to avoid vibration in self-locking drives, lead screw should have low axial and high torsional stiffness. Gallina [20] further expanded this study and, using both eigenvalue analysis and experiments, showed that by increasing lead screw moment of inertia it is possible to avoid instability under certain conditions.

Oledzki [21] studied self-locking mechanisms. He classified all types of mechanical drives, including worm gears and lead screws, with the emphasis on the possibility of self-locking. A unified notation was used to present geometrical features of the drives and to derive the equations of motion of a general kinematic pair. He also modeled the kinematic pair using elastic contacts instead of rigid contacts. The simulation results presented showed the possibility of "stick-slip" vibrations.

Generally, in high-accuracy linear positioning applications, "ball screws" are used because of their low friction, high lead accuracy, and backlash-free operation [22]. Consequently, the majority of works in the literature regarding position control and dynamics of screw drives focus on ball screws [23–30]. Lead screws are also used for similar positioning applications. For example, Otsuka [23] compared a high-precision lead screw drive equipped with an anti-backlash nut with two types of ball screw drives for nanometer positioning applications. The experimental results obtained showed the possibility of achieving nanometer accuracy with all three systems. Particular to the lead screw, the nonlinear behavior of the drive due to the stick-slip phenomenon was studied. The anti-backlash nuts were found to have an adverse effect due to preloading of the threads and increased friction.

Sato et al. [31] considered the dynamics of a lead screw positioning system with backlash. They set up an experiment using a sliding table, a lead screw, and a DC motor. In their experiments the table position, screw rotation angle, and DC motor current were measured. Although they did not undertake detailed modeling of lead screw and nut interaction, they were able to estimate lead screw/nut friction using a disturbance observer under the action of a linear proportional plus derivative feedback controller.

It is worth mentioning that lead screw drives were also used in redundant positioning systems for only coarse table motion [24, 25]. In these systems, a high-precision parallel positioning system such as a piezoelectric actuator is used for fine-tuning. Another example is the work by Sato et al. [32], where they introduced an active lead screw mechanism. By using two nuts connected together by a piezoelectric actuator, they were able to actively control backlash to achieve position accuracy of better than 10 nm.

1.3 Outline

Some of the geometrical aspects of lead screws are reviewed briefly in Chap. 2. This chapter also includes a short discussion of the various topics in *lead screw engineering* which highlights the need for the present study. In Chap. 3, some background materials are reviewed which cover the analytical tools we will use to analyze friction-induced instability in lead screw drives. The three mechanisms through which sliding friction can cause instability in a dynamical system are introduced in Chap. 4. For each of these three friction-induced instability mechanisms, simple examples are presented and analyzed in detail. The study of these mechanisms in the lead screw drives constitutes the main focus of this monograph. A series of lumped-mass lead screw drive models of varying number of degrees of freedom (DOFs) are developed in Chap. 5. These models are the basis of the analytical, numerical, and experimental studies presented in the later chapters. Chapters 6–8 are dedicated to the study of *negative damping instability mechanism*, *mode coupling instability mechanism*, and *kinematic constraint instability mechanisms* in lead screw drives, respectively. Finally in Chap. 9, an experimental case study is presented where the lead screw of the horizontal motion drive of a powered seat adjuster is examined. In this chapter, the aim is to address to excessive noise problem of the lead screw drive through analytical and experimental studies.

Chapter 2
Lead Screws

2.1 Screw Threads

The screw is the last machine to joint the ranks of the six fundamental simple machines. It has a history that stretches back to the ancient times. A very interesting historical account of the development of screws from Archimedes' water snail to the works of Leonardo da Vinci and up to the twentieth century is given by Mac Kenzie [33].

The mechanics of a screw is similar to two other simple machines, namely; the *inclined plane* and the *wedge*. As shown in Fig. 2.1, a screw can be considered as an inclined plane wrapped around a cylinder. Similar to the inclined plane, the horizontal force F needed to raise a weight W is

$$F \geq \left(\frac{\mu + \tan\lambda}{1 - \mu\tan\lambda}\right) W,$$

where μ is the coefficient of friction of the two rubbing surfaces and λ is the lead angle (equivalent to the angle that the inclined plane makes with the horizon).

Figure 2.2 compares a screw with a wedge. Here, instead of moving the load, the wedge is pushed under the load to raise it. The screw equivalent of this mechanism operates by applying a torque T to the screw to push the load upward turn by turn. Here the torque T needed to raise a weight W is

$$T \geq r\left(\frac{\mu + \tan\lambda}{1 - \mu\tan\lambda}\right) W.$$

The above force mechanisms are shared by both fastening screws and translating screws. The screws in the latter group – studied in this monograph – are commonly known as lead screws and are used for transmitting force and/or positioning by converting rotary to translational motion. In power transmission applications, lead screws are also known as "power screws" [34, 35]. When used in vertical applications, these systems are sometimes called "screw jacks" [1].

O. Vahid-Araghi and F. Golnaraghi, *Friction-Induced Vibration in Lead Screw Drives*,
DOI 10.1007/978-1-4419-1752-2_2,

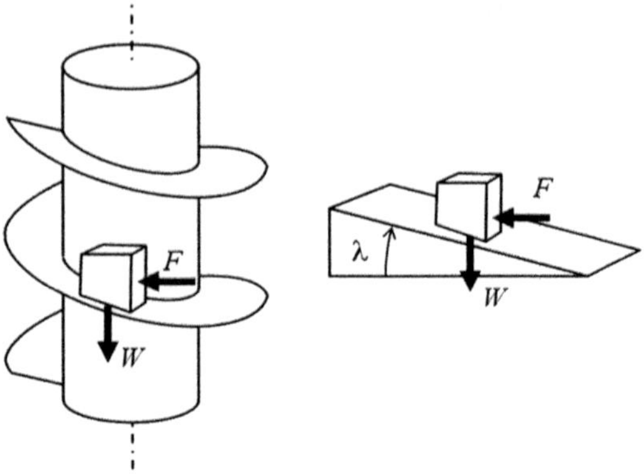

Fig. 2.1 Screw as an inclined plane wrapped around a cylinder

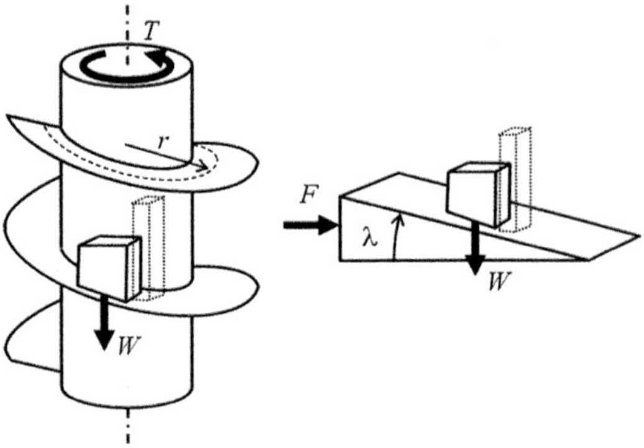

Fig. 2.2 Screw as a wedge wrapped around a cylinder

There are a number of thread geometries available for lead screws that are designed to address various requirements such as ease of manufacturing, load-carrying capacity, and the quality of fit [33]. The most popular of these geometries are the Acme and stub-Acme threads.[1] Figure 2.3 shows the basic dimensions of symmetric trapezoidal threads (e.g., Acme threads). The thread angle (ψ_a) for Acme and stub Acme thread is 14½°. The basic relationships defining the screw geometry are given next for future reference.[2] The lead angle (or helix angle), λ, is defined as

$$\tan \lambda = \frac{l}{\pi d_{\mathrm{m}}}, \tag{2.1}$$

[1]This design is further discussed in Sect. 2.3.

[2]See [33] for specifications of other types of screw threads.

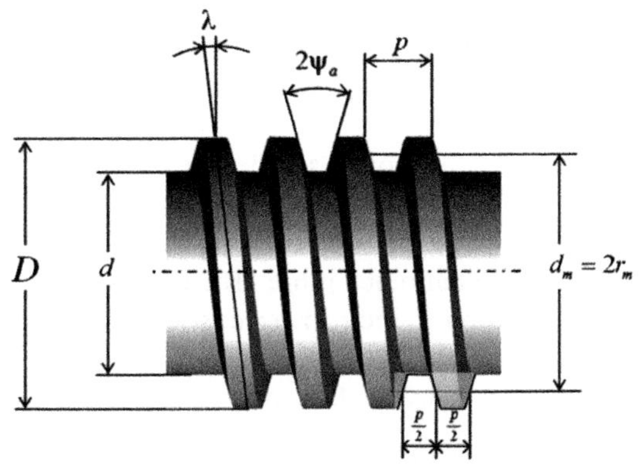

Fig. 2.3 Dimensions of screw threads

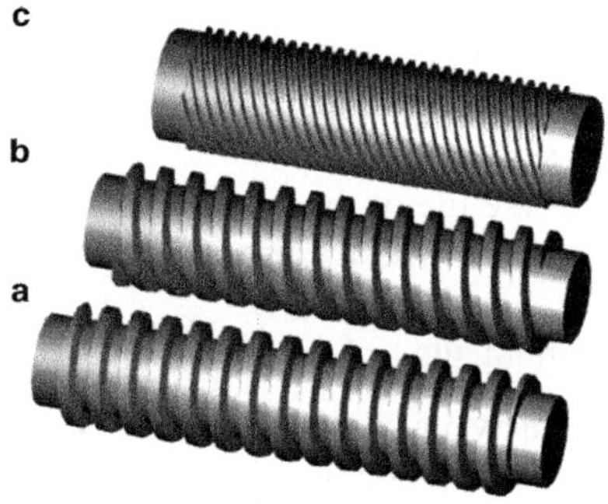

Sample	Nom. Dia.	Pitch	Starts	Lead
A	1"	0.25"	1	0.25"
B	1"	0.25"	2	0.50"
C	1"	0.10"	10	1.00"

Fig. 2.4 Screws with various numbers of starts

where d_m is the *pitch diameter* and l is the *lead* and it is defined as

$$l = n_s \cdot p, \tag{2.2}$$

where p is the screw *pitch* (distance between identical points of two consecutive threads) and n_s is the number of starts (or *starts*). Figure 2.4 shows three 1-in. lead screws with one, two, and ten starts.

Increasing the number of starts increases the lead thus increasing the translational velocity of the nut for a given fixed angular velocity of the screw. Based on (2.1) and (2.2) the lead angles for these screws are found as follows: $\lambda_{(A)} \cong 5.20°$, $\lambda_{(B)} \cong 10.31°$, and $\lambda_{(C)} \cong 18.52°$. In these examples, the pitch diameter was found according to the following equation: $d_m = D - (p/2) = d + (p/2)$.

2.2 Lead Screw Engineering

For design and selection purposes, the mechanical analysis of lead screws usually is limited to the factors affecting their static or quasi-static performance, such as efficiency, driving torque requirements, and load capacity [33–35]. There are numerous important aspects involved in the successful design of a lead screw drive system. Some of these issues are summarized in Fig. 2.5. It is important to mention that, to some degree, almost all of these issues influence the other aspects of the lead screw design.

Manufacturers offer a wide range of products in response to the diverse applications where lead screws are utilized. For positioning stages, high precision ground lead screws with or without anti-backlash nuts are offered as an alternative to the more costly but much more efficient ball screw-driven stages [36, 37].

In addition to their lower cost compared to ball screws, there are a number of distinct features that make a lead screw drive the favorable choice – if not the only choice – in many applications. These features include [38–40] the following:

- Quieter operation due to the absence of re-circulating balls used in ball screws.
- Smaller moving mass and smaller packaging.
- Availability of high helix angles resulting in very fast leads.
- Availability of very fine threads for high resolution applications.
- Possibility of self-locking to prevent the drive from being backdrivable thus eliminating the need for a separate brake system.
- Lower average particulate generation over the life of the system.
- Elimination of the need for periodic lubrication with the use of self-lubricating polymer nuts.
- Possibility to work in washed-down environments.

Design factors given in Fig. 2.5 are discussed by the manufacturers as part of their public technical information or product selection guidelines (see, e.g., [41–46]). There is, however, a major exception: friction-induced vibration. Only a few published works are found in the literature that discuss the dynamics of lead screw drive systems and the effect of friction on their vibratory behavior.[3]

Wherever sliding motion exists in machines and mechanisms, friction-induced vibration may occur, and when it does, it severely affects the function of the system. Excessive noise, diminished accuracy, and reduced life are some of the adverse consequences of friction-induced vibration. To this end, lead screw systems are no exception; the lead screw threads slide against meshing nut threads as the system operates.

One of the common issues in using lead screws – especially for the positioning applications – is backlash. As shown in Fig. 2.6, backlash is the axial distance the nut can be moved without turning the lead screw. Among the problems caused by backlash are the deterioration of the positioning accuracy and diminished repeatability of the performed task by the lead screw drive. Both design and/or manufacturing factors may contribute to the presence of backlash in a lead screw drive. Various anti-backlash nuts are designed and offered by the manufacturers to address these problems. These nuts generally are made of two halves connected

[3]See Sect. 1.2.

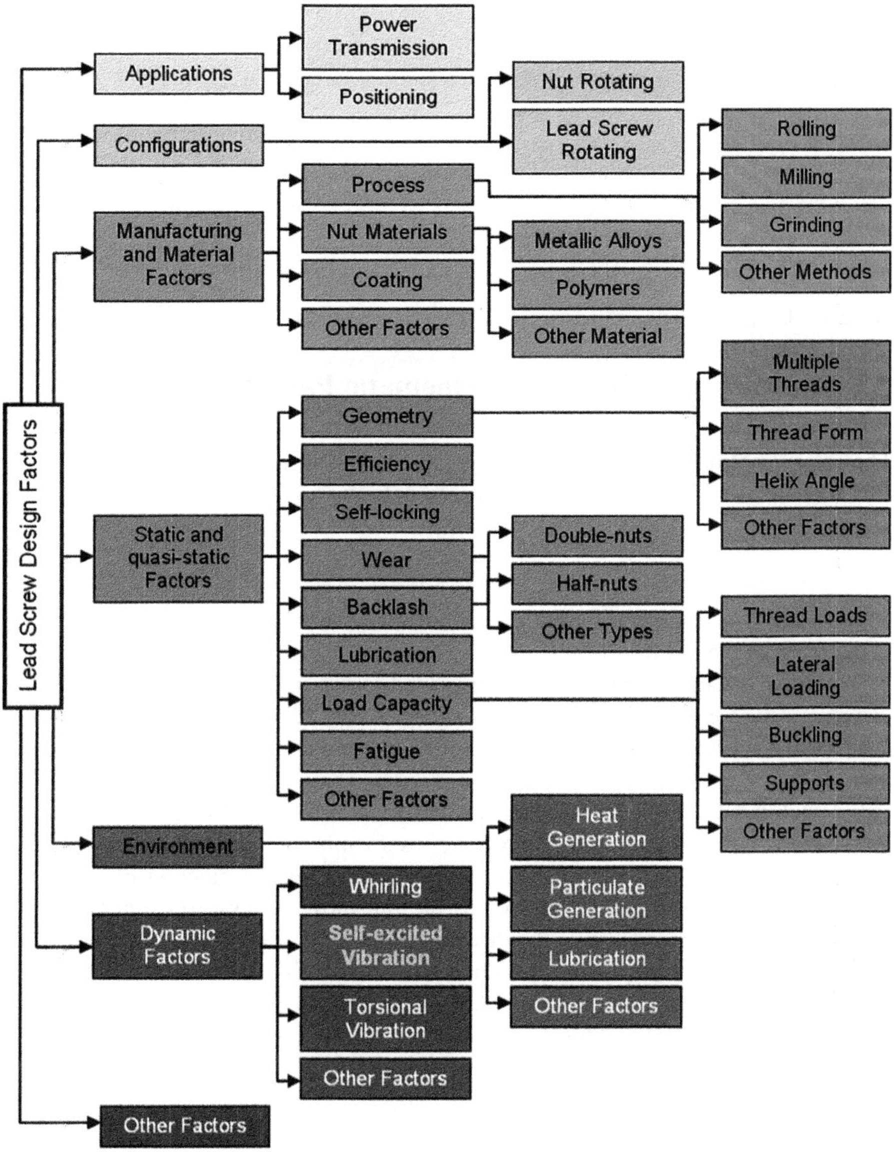

Fig. 2.5 Lead screw design and selection factors

with preloaded springs that can move with respect to one another to compensate backlash and wear [36, 41–44]. The drawback of using these nuts is in the increased friction force, which lowers the efficiency and increases the required driving torque.[4]

[4]See Sect. 5.4 for a mathematical model of a lead screw with an anti-backlash nut.

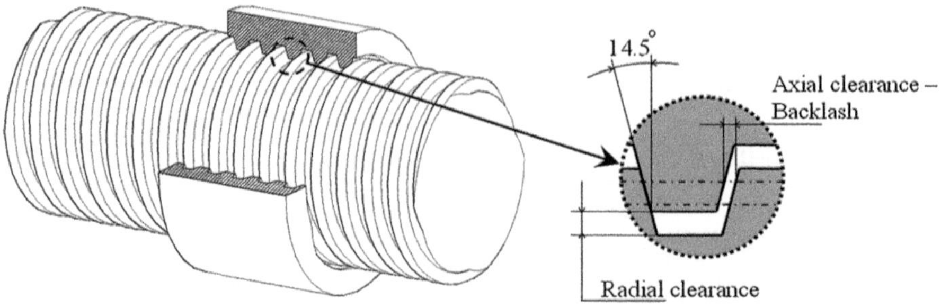

Fig. 2.6 Meshing "stub Acme" lead screw and nut (*cut view*). Detail: radial and axial clearances

2.3 Lead Screw and Nut: A Kinematic Pair

The rotary motion is converted to linear translation at the interface of lead screw and nut threads. The kinematic relationship defining a lead screw is simply[5]

$$x = r_{\mathrm{m}} \tan \lambda \theta, \tag{2.3}$$

where θ is the lead screw rotation, x is the nut translation, λ is the lead angle, and r_{m} is the pitch circle radius.

The interaction between the contacting lead screw and nut threads can be easily visualized by considering unrolled threads (see Figs. 2.1 and 2.2). This way, the rotation of lead screw is replaced by an equivalent translation. Assuming one thread pair to be in contact at any given instant, Fig. 2.7 shows the interaction of the lead screw and nut threads for both left-handed and right-handed screws. The sign conventions used for the contact force, N, is shown in this figure. In the configurations shown, when the right-handed lead screw is rotated clockwise/moved up (rotated counterclockwise/moved down) the nut moves backward/right (forward/left). For the left-handed screw, the direction of motion of the nut is reversed. Also, when the nut threads are in contact with the leading (trailing) lead screw threads, the normal component of contact force, N, is considered to be positive (negative).

The friction force is given by

$$F_{\mathrm{f}} = \mu |N| \mathrm{sgn}(v_{\mathrm{s}}), \tag{2.4}$$

where μ is the coefficient of friction (possibly velocity dependent) and v_{s} is the relative sliding velocity. The friction force acts tangent to the contacting thread surfaces and always opposes the direction of motion but does not change direction when normal force, N, changes direction.

[5] By properly orienting the x-axis, this relationship applies to both left-hand and right-hand threads.

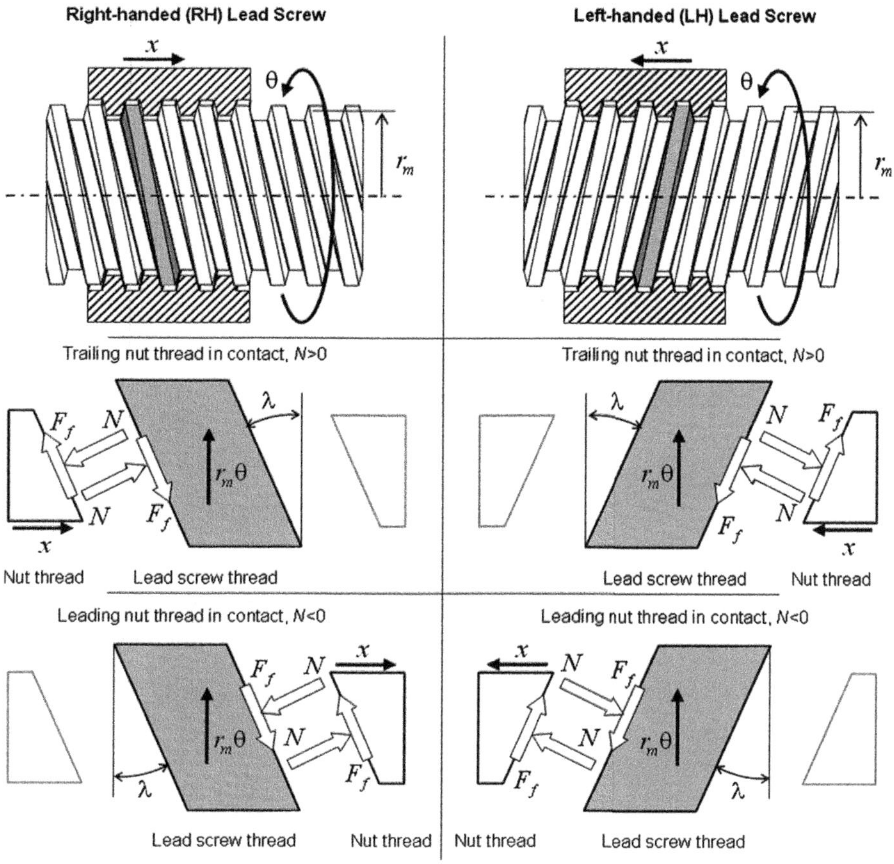

Fig. 2.7 Sign convention for contact forces between nut and lead screw

2.4 Effect of Thread Angle

Before moving on to the dynamic models of lead screw systems, the effect of thread geometry on the contact forces is considered here. The force interaction shown in Fig. 2.7 is essentially correct for the square threads where the normal force is parallel to the lead screw axis. For Acme or other types of threads, a slight modification is needed to take into account the thread angle.

Figure 2.8 shows the thread semi-angles as measured on a section through the axis of a screw, ψ_a, and as measured on a section perpendicular to the helix, ψ_n.

Using the geometric relationship in Fig. 2.9, one can write [47]

$$\tan\psi_n = \frac{x_n}{y}, \quad \tan\psi_a = \frac{x_a}{y}, \tag{2.5}$$

$$x_n = x_a \cos\lambda. \tag{2.6}$$

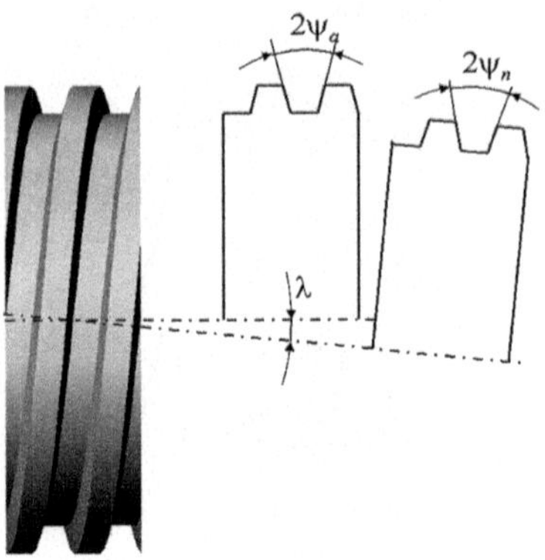

Fig. 2.8 Effect of lead angle on the measurement of thread angle

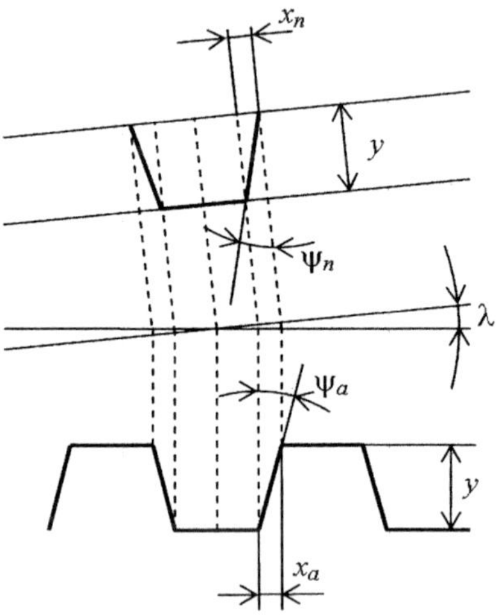

Fig. 2.9 Geometry of the threads on two different section planes

Combining (2.5) and (2.6) gives

$$\tan\psi_n = \tan\psi_a \cos\lambda.$$

Figure 2.10 shows a portion of a lead screw with localized contact force $\hat{N}$ (perpendicular to the thread surface) and friction force F_f. The X-axis of XYZ coordinate system is parallel to the lead screw axis. The x-Z plane is perpendicular to the helix. The projection of contact force on the x-y (or X-Y) plane is calculated as

$$N = \hat{N}\cos\psi_n. \tag{2.7}$$

Since $\hat{N}$ is the normal force, using (2.4) the friction force for trapezoid threads is calculated by $F_f = \hat{\mu}|\hat{N}|\text{sgn}(\dot{\theta})$, where $\hat{\mu}$ is the true coefficient of friction. One can define the *apparent* coefficient of friction as

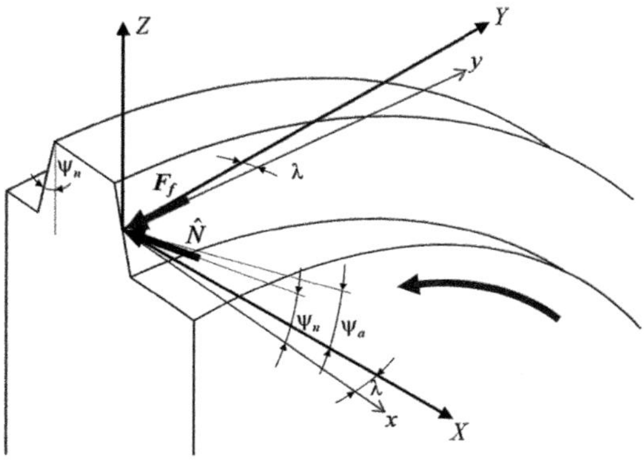

Fig. 2.10 Forces acting on a thread

$$\mu = \hat{\mu}\cos\psi_n = \frac{\mu}{\sqrt{\tan^2\psi_a\cos^2\lambda + 1}}. \tag{2.8}$$

Using (2.7) and (2.8), the friction force is written conveniently as $F_f = \mu|N|\text{sgn}(\dot{\theta})$, which is the same as (2.4) and will be used in the subsequent chapters.

Chapter 3
Some Background Material

The study of the stability of linear dynamical systems is a well-established field and there are many references available in the literature on this subject (see, e.g., [48–50]). In this chapter, a brief review of the relevant material to the subsequent study of the dynamical systems with friction is presented. An important tool in the study of the stability of mechanical systems is the eigenvalue analysis method. The *local* stability of the system's equilibrium point is determined by studying the sign of the real part of the eigenvalues of the system's Jacobian matrix. As parameters are changed, one or more eigenvalues may cross the imaginary axis marking the onset of the instability.

The linearized equations of motion of the dynamical systems are presented in Sect. 3.1. An introduction to the modeling of dynamical systems that include frictional constraints is given in Sect. 3.2. In Sect. 3.3, a classification of the linearized equations of motion is given and the consequences of the nonconservative forces such as friction is discussed. The eigenvalue stability analysis method is reviewed briefly in Sects. 3.4 and 3.5 for the general case and the undamped case, respectively.

The method of first-order averaging is introduced in Sect. 3.6. This method is utilized in Sect. 4.1 and Chap. 6 to expand the results of the eigenvalue analysis in the study of negative damping instability mechanism.

3.1 Linearized System Equations

Consider the following second-order nonlinear autonomous system:

$$\ddot{\mathbf{x}} = \mathbf{f}(\mathbf{x}, \dot{\mathbf{x}}), \tag{3.1}$$

where $\mathbf{f} : R^n \to R^n$ is a vector function and $\mathbf{f}(\mathbf{0}, \mathbf{0}) = 0$. Also $\mathbf{x}$ is a column vector of n variables quantifying the n degrees-of-freedom (DOFs) of the dynamical system. Define a $2n$-vector of system states as $\mathbf{y} = \begin{bmatrix} \mathbf{x}^{\mathrm{T}} & \dot{\mathbf{x}}^{\mathrm{T}} \end{bmatrix}^{\mathrm{T}}$. The second-order system (3.1) can be rewritten in state-space form as

O. Vahid-Araghi and F. Golnaraghi, *Friction-Induced Vibration in Lead Screw Drives*, DOI 10.1007/978-1-4419-1752-2_3, © Springer Science+Business Media, LLC 2011

$$\dot{\mathbf{y}} = \frac{\mathrm{d}}{\mathrm{d}t}\begin{bmatrix}\mathbf{x}\\ \dot{\mathbf{x}}\end{bmatrix} = \begin{bmatrix}\frac{\mathrm{d}}{\mathrm{d}t}\mathbf{x}\\ \mathbf{f}(\mathbf{x},\dot{\mathbf{x}})\end{bmatrix} \equiv \mathbf{F}(\mathbf{y}), \tag{3.2}$$

where $\mathbf{F}(\mathbf{0}) = 0$. The linearized system in a neighborhood of the equilibrium point $\mathbf{y} = 0$ can be written as

$$\dot{\mathbf{y}} = \mathbf{A}\ \mathbf{y}, \tag{3.3}$$

where (assuming that the partial derivatives exist),

$$\mathbf{A} = \left.\frac{\partial \mathbf{F}}{\partial \mathbf{y}}\right|_{\mathbf{y}=0} \tag{3.4}$$

is the system's Jacobian matrix evaluated at the origin.

In many dynamical systems the equation of motion of the system takes the form

$$\widehat{\mathbf{M}}(\mathbf{x})\ \ddot{\mathbf{x}} + \mathbf{h}(\mathbf{x},\dot{\mathbf{x}}) = 0, \tag{3.5}$$

where $\mathbf{h}(\mathbf{0},\mathbf{0}) = 0$ (i.e., the origin is an equilibrium point of the system) and $\widehat{\mathbf{M}}(\mathbf{x})$ is symmetric and positive definite for all $\mathbf{x}$. Here we assume that the elements of matrix $\widehat{\mathbf{M}}(\mathbf{x})$ and vector $\mathbf{h}(\mathbf{x},\dot{\mathbf{x}})$ are sufficiently smooth functions of their arguments in a neighborhood of the origin. Compared to (3.1), we have

$$\mathbf{f}(\mathbf{x},\dot{\mathbf{x}}) = -\widehat{\mathbf{M}}^{-1}(\mathbf{x})\ \mathbf{h}(\mathbf{x},\dot{\mathbf{x}}).$$

Expanding $\mathbf{h}(\mathbf{x},\dot{\mathbf{x}})$, one can write

$$\mathbf{h}(\mathbf{x},\dot{\mathbf{x}}) = \Lambda_p\ \mathbf{x} + \Lambda_v\ \dot{\mathbf{x}} + O\left(\ \|\mathbf{x}\|^2\right) + O\left(\ \|\dot{\mathbf{x}}\|^2\right), \tag{3.6}$$

where

$$\Lambda_p = \left.\frac{\partial \mathbf{h}}{\partial \mathbf{x}}\right|_{(\mathbf{x},\dot{\mathbf{x}})=(\mathbf{0},\mathbf{0})} \quad \text{and} \quad \Lambda_v = \left.\frac{\partial \mathbf{h}}{\partial \dot{\mathbf{x}}}\right|_{(\mathbf{x},\dot{\mathbf{x}})=(\mathbf{0},\mathbf{0})}.$$

Also,

$$\widehat{\mathbf{M}}^{-1}(\mathbf{x}) = \mathbf{M}^{-1} + O(\ \|\mathbf{x}\|) + O(\ \|\dot{\mathbf{x}}\|), \tag{3.7}$$

where $\mathbf{M} = \widehat{\mathbf{M}}(\mathbf{0})$. Substituting (3.6) and (3.7) into (3.5) and discarding the *nonlinear* terms, one finds the linearized system equation as

$$\mathbf{M}\ \ddot{\mathbf{x}} + \Lambda_v\ \dot{\mathbf{x}} + \Lambda_p\ \mathbf{x} = \mathbf{0}. \tag{3.8}$$

For this system, (3.4) takes the form

$$\mathbf{A} = \begin{bmatrix} \mathbf{0}_{n\times n} & \mathbf{1}_{n\times n} \\ -\mathbf{M}^{-1}\Lambda_p & -\mathbf{M}^{-1}\Lambda_v \end{bmatrix}, \tag{3.9}$$

where $\mathbf{0}_{n\times n}$ and $\mathbf{1}_{n\times n}$ are $n \times n$ zero and identity matrices, respectively.

3.2 Equations of Motion with Contact Forces

Consider an n-DOF dynamical system with a single frictional contact. The equation of motion of the systems in the matrix form can be written as [51, 52]

$$\widehat{\mathbf{M}}(\mathbf{x})\ddot{\mathbf{x}} + \mathbf{H}(\mathbf{x}, \dot{\mathbf{x}}) = N\,\mathbf{v}_n - F_\mathrm{f}\mathbf{v}_t, \tag{3.10}$$

where $\mathbf{x} \in R^n$ is the vector of the generalized coordinates, N and F_f are the normal and tangential (friction) forces of the contact acting in $\mathbf{v}_n \in R^n$ and $\mathbf{v}_t \in R^n$ directions, respectively. Also, $\widehat{\mathbf{M}}$ is a symmetric and positive definite matrix and $\mathbf{H}$ is the vector of smooth generalized forces. For simplicity, we assume that the constraint equation defining the (bilateral) contact is linear in the generalized coordinates, i.e.,

$$g = \mathbf{w}^\mathrm{T} \cdot \mathbf{x} = 0. \tag{3.11}$$

Note that from the above definitions we have; $\mathbf{w} = \alpha\mathbf{v}_n$ for some constant $\alpha \neq 0$. Let $\mathbf{y} \in R^{n-1}$ be the vector of *reduced* generalized coordinates. From (3.11), we have

$$\mathbf{x} = \mathbf{Q}\mathbf{y}, \tag{3.12}$$

where $\mathbf{w}^\mathrm{T}\mathbf{Q} = 0$. Substituting (3.12) into (3.10) and multiplying the result by $\mathbf{Q}^\mathrm{T}$ yield

$$\mathbf{Q}^\mathrm{T}\widehat{\mathbf{M}}\mathbf{Q}\ddot{\mathbf{y}} + \mathbf{Q}^\mathrm{T}\mathbf{H} = -\mu_s N\mathbf{Q}^\mathrm{T}\mathbf{v}_t, \ddot{\mathbf{y}} \tag{3.13}$$

where we used the fact that $\mathbf{Q}^\mathrm{T}\mathbf{v}_n = 0$ and also the friction force is written as

$$F_\mathrm{f} = \mu_s N = \mu N \mathrm{sgn}(v_t)\mathrm{sgn}(N), \tag{3.14}$$

where $v_t = v_t(\mathbf{x}, \dot{\mathbf{x}})$ is the contact sliding velocity and μ is the coefficient of friction. Substituting (3.14) into (3.10) and solving for $\ddot{\mathbf{x}}$ yield

$$\ddot{\mathbf{x}} = \widehat{\mathbf{M}}^{-1}[-\mathbf{H} + N(\mathbf{v}_n - \mu_s\mathbf{v}_t)].$$

Now, multiplying both sides of this equation by $\mathbf{w}^{\mathrm{T}}$ and solving for N give

$$AN = b, \tag{3.15}$$

where (3.11) was used and

$$A(\mathbf{x}, \dot{\mathbf{x}}) = \mathbf{w}^{\mathrm{T}} \widehat{\mathbf{M}}^{-1} (\mathbf{v}_n - \mu_s \mathbf{v}_t), \tag{3.16}$$

$$b(\mathbf{x}, \dot{\mathbf{x}}) = \mathbf{w}^{\mathrm{T}} \widehat{\mathbf{M}}^{-1} \mathbf{H}. \tag{3.17}$$

Multiplying (3.13) by $\mathbf{A}$ and substituting (3.15) yield

$$\tilde{\mathbf{M}}(\mathbf{y}, \dot{\mathbf{y}})\ddot{\mathbf{y}} + \tilde{\mathbf{H}}(\mathbf{y}, \dot{\mathbf{y}}) = 0, \tag{3.18}$$

where

$$\tilde{\mathbf{M}}(\mathbf{y}, \dot{\mathbf{y}}) = A\mathbf{Q}^{\mathrm{T}} \widehat{\mathbf{M}} \mathbf{Q}, \tag{3.19}$$

$$\tilde{\mathbf{H}}(\mathbf{y}, \dot{\mathbf{y}}) = A\mathbf{Q}^{\mathrm{T}} \mathbf{H} + \mu_s b \mathbf{Q}^{\mathrm{T}} \mathbf{v}_t, \tag{3.20}$$

where the substitution $\mathbf{x} = \mathbf{Q}\mathbf{y}$ must be applied to the arguments of the functions on the right-hand-side of (3.19) and (3.20).

The reduced system (3.18) takes the form of (3.5). There is, however, one very important difference between the two systems; the inertia matrix given by (3.19) is zero whenever A vanishes.

Remark 3.1 For the case of a unilateral contact, (3.14) must be changed to

$$F_{\mathrm{f}} = \mu_s N = \mu N \mathrm{sgn}(v_t),$$

and the above formulation is used to study the system's behavior while the two contacting bodies are in contact, i.e., $\dot{g} = \mathbf{w}^{\mathrm{T}} \cdot \dot{\mathbf{x}} = 0$ and $N \geq 0$. □

Remark 3.2 The above formulation can be extended to cases where multiple contacting pairs with friction exist. See, for example, [51–53]. □

Remark 3.3 An equivalent form of the reduced-DOF system may be written with an *asymmetric* inertia matrix (see, e.g., Sects. 5.6 and 5.8). In this case, parameters satisfying $A = 0$ result in a singular inertia matrix. □

The situations where $A \leq 0$ are known as Painlevé's paradoxes [51]. The paradoxes arise from the violation of the existence and uniqueness conditions of the solution of the system's equations of motion, (3.18). We'll discuss these cases further in Sect. 4.3.

3.3 Classification of Linear Systems

The position and velocity coefficient matrices (i.e., $\mathbf{\Lambda}_p$ and $\mathbf{\Lambda}_v$) in (3.8) are, in general, not symmetric and represent both conservative and nonconservative forces. We consider the following types of autonomous general forces [48]:

1. Velocity-independent forces: $\mathbf{f}(\mathbf{x})$
2. Velocity-dependent forces: $\mathbf{f}(\mathbf{x}, \dot{\mathbf{x}})$

The first category includes conservative forces that contribute the term $\mathbf{K}_1\,\mathbf{x}$, where $\mathbf{K}_1 = \mathbf{K}_1{}^{\mathrm{T}} > 0$ to the linearized equations of motion. This category also includes the circulatory or follower forces that contribute the asymmetric term $(\mathbf{K}_2 + \mathbf{S}_2)\,\mathbf{x}$, where $\mathbf{K}_2 = \mathbf{K}_2{}^{\mathrm{T}}$ and $\mathbf{S}_2 = -\mathbf{S}_2{}^{\mathrm{T}}$ (i.e., matrix $\mathbf{S}_2$ is skew-symmetric). The second category includes dissipative forces (i.e., damping) which contribute the term $\mathbf{C}_1\,\dot{\mathbf{x}}$, where $\mathbf{C}_1 = \mathbf{C}_1{}^{\mathrm{T}} \geq 0$. Conservative gyroscopic forces encountered in rotating systems belong to this category and contribute the term $\mathbf{G}\,\dot{\mathbf{x}}$, where $\mathbf{G} = -\mathbf{G}^{\mathrm{T}}$. Finally there are cases where forces depend on both position and velocity and can be represented (in linearized form) by $\mathbf{S}_3\,\mathbf{x} + \mathbf{C}_3\,\dot{\mathbf{x}}$, where $\mathbf{S}_3 = -\mathbf{S}_3{}^{\mathrm{T}}$ and $\mathbf{C}_3 = \mathbf{C}_3{}^{\mathrm{T}}$ [48].

In the general case where all of the above forces are present, the linearized equations of motion of the n-DOF autonomous mechanical system takes the form

$$\mathbf{M}\ddot{\mathbf{x}} + (\mathbf{C} + \mathbf{G})\,\dot{\mathbf{x}} + (\mathbf{K} + \mathbf{S})\,\mathbf{x} = 0, \tag{3.21}$$

where $\mathbf{M}$ is the symmetric positive definite inertia matrix, $\mathbf{C} = \mathbf{C}_1 + \mathbf{C}_3$ is the symmetric matrix of damping coefficients, $\mathbf{G}$ is the skew-symmetric matrix representing the gyroscopic forces, $\mathbf{K} = \mathbf{K}_1 + \mathbf{K}_2$ is the symmetric stiffness matrix, and $\mathbf{S} = \mathbf{S}_2 + \mathbf{S}_3$ is the skew-symmetric matrix representing the nonconservative forces.

Remark 3.4 The relationship between general matrix coefficients $\mathbf{\Lambda}_p$ and $\mathbf{\Lambda}_v$ in (3.8) and the matrices in (3.21) can be written as

$$\mathbf{K} = \frac{1}{2}\left(\mathbf{\Lambda}_p + \mathbf{\Lambda}_p^{\mathrm{T}}\right), \quad \mathbf{S} = \frac{1}{2}\left(\mathbf{\Lambda}_p - \mathbf{\Lambda}_p^{\mathrm{T}}\right)$$

and

$$\mathbf{C} = \frac{1}{2}\left(\mathbf{\Lambda}_v + \mathbf{\Lambda}_v^{\mathrm{T}}\right), \quad \mathbf{G} = \frac{1}{2}\left(\mathbf{\Lambda}_v - \mathbf{\Lambda}_v^{\mathrm{T}}\right).$$

□

Based on (3.21), the following general classification of linear systems can be made:

- Conservative nongyroscopic systems:

$$\mathbf{M}\ddot{\mathbf{x}} + \mathbf{K}\,\mathbf{x} = 0.$$

- Conservative gyroscopic systems:

$$\mathbf{M}\ddot{\mathbf{x}} + \mathbf{G}\,\dot{\mathbf{x}} + \mathbf{K}\,\mathbf{x} = 0.$$

- Damped linear nongyroscopic systems:

$$\mathbf{M}\ddot{\mathbf{x}} + \mathbf{C}\,\dot{\mathbf{x}} + \mathbf{K}\,\mathbf{x} = 0.$$

- Damped gyroscopic systems:

$$\mathbf{M}\ddot{\mathbf{x}} + (\mathbf{C} + \mathbf{G})\,\dot{\mathbf{x}} + \mathbf{K}\,\mathbf{x} = 0.$$

- Undamped Circulatory systems:

$$\mathbf{M}\ddot{\mathbf{x}} + (\mathbf{K} + \mathbf{S})\,\mathbf{x} = 0.$$

- Damped circulatory systems:

$$\mathbf{M}\ddot{\mathbf{x}} + (\mathbf{C} + \mathbf{G})\,\dot{\mathbf{x}} + (\mathbf{K} + \mathbf{S})\,\mathbf{x} = 0.$$

Specific to the case of dynamic systems with frictional constraints, two other subcategories of circulatory systems can be included in the above list where the inertia matrix, $\tilde{\mathbf{M}}$, is asymmetric due to friction:

- Undamped Circulatory systems:

$$\tilde{\mathbf{M}}\ddot{\mathbf{y}} + \mathbf{K}\,\mathbf{y} = 0. \tag{3.22}$$

- Damped circulatory systems:

$$\tilde{\mathbf{M}}\ddot{\mathbf{y}} + \mathbf{C}\,\dot{\mathbf{y}} + \mathbf{K}\,\mathbf{y} = 0. \tag{3.23}$$

- *Remark 3.5* Systems given by (3.22) and (3.23) can be converted to the regular form of circulatory systems with a symmetric positive definite inertia matrix and asymmetric stiffness and damping matrices. First notice that in the absence of friction the inertia matrix *is* symmetric and positive definite.[1] Let $\mathbf{M} = \tilde{\mathbf{M}}|_{\mu=0}$. If $\det(\tilde{\mathbf{M}}) \neq 0$, multiplying (3.23) by the matrix $\mathbf{M} \cdot \tilde{\mathbf{M}}^{-1}$ yields

$$\mathbf{M}\,\ddot{\mathbf{y}} + \boldsymbol{\Lambda}_v\dot{\mathbf{y}} + \boldsymbol{\Lambda}_p\mathbf{y} = 0,$$

where $\boldsymbol{\Lambda}_p = \mathbf{M} \cdot \tilde{\mathbf{M}}^{-1} \cdot \mathbf{K}$ and $\boldsymbol{\Lambda}_v = \mathbf{M} \cdot \tilde{\mathbf{M}}^{-1} \cdot \mathbf{C}$. In general, $\boldsymbol{\Lambda}_p$ and $\boldsymbol{\Lambda}_v$ are asymmetric. □

- *Remark 3.6* In line with the discussions of Sect. 3.2, an alternative but equivalent formulation suitable for the cases where the inertia matrix may become singular in the parameter range of interest is

$$\det(\tilde{\mathbf{M}})\,\mathbf{M}\,\ddot{\mathbf{y}} + \mathbf{M}\,\mathrm{adj}\tilde{\mathbf{M}}\,\mathbf{C}\,\dot{\mathbf{y}} + \mathbf{M}\,\mathrm{adj}\tilde{\mathbf{M}}\,\mathbf{K}\,\mathbf{y} = 0, \tag{3.24}$$

where $\mathrm{adj}\tilde{\mathbf{M}}$ is the adjoint of $\tilde{\mathbf{M}}$. □

3.4 Stability Analysis

The following two definitions are necessary to clarify the subsequent study of the linear stability in dynamical systems.

Definition 3.1. *(Lyapunov Stability)* Consider the autonomous system (3.2) in a neighborhood $D \subset R^n$ of $\mathbf{y} = 0$. The equilibrium point $\mathbf{y} = 0$ is called stable (in the sense of Lyapunov) if for each $\varepsilon > 0$ there exists $\delta(\varepsilon) > 0$ such that $\|\mathbf{y}(0)\| \le \delta$ yields $\|\mathbf{y}(t)\| \le \varepsilon$ for all $t \ge 0$. The equilibrium point is unstable, otherwise. □

Definition 3.2. *(Asymptotic Stability)* Consider the autonomous system (3.2) in a neighborhood $D \subset R^n$ of $\mathbf{y} = 0$. The equilibrium point $\mathbf{y} = 0$ is called asymptotically stable if it is stable and $\delta > 0$ can be chosen such that $\|\mathbf{y}(0)\| \le \delta$ yields $\lim_{t\to\infty} \mathbf{y}(t) = 0$. □

Let $\mathbf{x}(t) = \mathbf{a}e^{\eta t}$ be the solution of (3.8) where $\mathbf{a}$ is a constant n-vector and η is a parameter to be determined. Substituting this solution into (3.8) and simplifying gives

$$\left(\eta^2\mathbf{M} + \eta\boldsymbol{\Lambda}_v + \boldsymbol{\Lambda}_p\right) \cdot \mathbf{a} = 0, \tag{3.25}$$

which is equivalent to the more familiar form

$$\left(\mathbf{A} - \eta\mathbf{1}_{2n\times 2n}\right) \cdot \mathbf{b} = \mathbf{0}_{2n\times 1},$$

[1] In the absence of friction the kinematic constraint describing the contact is ideal and the inertia matrix of the reduced system model retains the symmetry and positive definiteness of the original system before applying the constraint equation.

where $\mathbf{b} = \begin{bmatrix} \mathbf{a}^{\mathrm{T}} & \eta\mathbf{a}^{\mathrm{T}} \end{bmatrix}^{\mathrm{T}}$ and $\mathbf{A}$ is given by (3.9).

For (3.25) to have nontrivial solutions (i.e., $\mathbf{a} \neq 0$), the matrix $\eta^2\mathbf{M} + \eta\Lambda_v + \Lambda_p$ must be singular. Thus, the characteristic equation is obtained as

$$\Delta(\eta) \equiv \det\left(\eta^2\mathbf{M} + \eta\Lambda_v + \Lambda_p\right) = 0$$

or

$$\Delta(\eta) = \det(\mathbf{M})\eta^{2n} + a_1\eta^{2n-1} + \cdots + a_{2n-1}\eta + \det\left(\Lambda_p\right) = 0, \tag{3.26}$$

which has $2n$ roots or *eigenvalues*, i.e., $\eta = \eta_j$, $j = 1 \ldots 2n$. Eigenvalues can be real or complex numbers. Since all of the coefficients in the eigenvalue problem (3.25) are real numbers, the complex eigenvalues occur in conjugate pairs. The origin is an *asymptotically stable* equilibrium point of (3.8) when all of the eigenvalues have *negative* real parts. Moreover, if the real part of at least one eigenvalue is positive, the origin is unstable. In the case where some eigenvalues have zero real parts and all other eigenvalues have negative real parts the origin is *stable* if and only if the Jordan blocks corresponding to eigenvalues with zero real parts are scalar blocks [54].

Assuming the general complex-valued eigenvalue as $\eta_j = \rho_j + i\omega_j$, where ρ_j and ω_j are real numbers, the following two types of instability are identified [48]:

- Flutter instability: $\exists j$ such that $\rho_j > 0$ and $\omega_j \neq 0$.
- Divergence instability: $\exists j$ such that $\rho_j > 0$ and $\omega_j = 0$.

The divergence is a statical instability where the system response grows exponentially. Flutter, on the other hand, is a dynamical instability and involves system vibration with growing amplitude [48].

Consider the case where one or more of the parameters of system (3.8) are varied. The eigenvalues are generally found as functions of these parameters, i.e., $\eta_j = \eta_j(\boldsymbol{\theta})$, where $\boldsymbol{\theta} = \begin{bmatrix} \theta_1 & \theta_2 & \cdots & \theta_k \end{bmatrix}^{\mathrm{T}}$ is the vector of system parameters. Assume that initially the origin is stable. Further assume that by varying the parameters, an eigenvalue (say *j-th* eigenvalue) crosses the imaginary axis at $\boldsymbol{\theta} = \boldsymbol{\theta}_{\mathrm{cr}}$, i.e., $\rho_j(\boldsymbol{\theta}_{\mathrm{cr}}) = 0$. If $\omega_j(\boldsymbol{\theta}_{\mathrm{cr}}) = 0$, then the surface $\rho_j(\boldsymbol{\theta}) = 0$ defines the divergence instability boundary in k-dimensional parameter space. On the other hand, if $\omega_j(\boldsymbol{\theta}_{\mathrm{cr}}) \neq 0$, this surface defines the flutter boundary. For the case of only one parameter, Figs. 3.1 and 3.2 show the evolution of an eigenvalue (or a pair of complex-conjugate eigenvalues) in the $\rho - \omega - \theta$ space for flutter and divergence cases, respectively.

From (3.26) it is easy to see that, the divergence boundary can be found by solving

$$\det\left[\Lambda_p(\boldsymbol{\theta}_{\mathrm{cr}})\right] = 0.$$

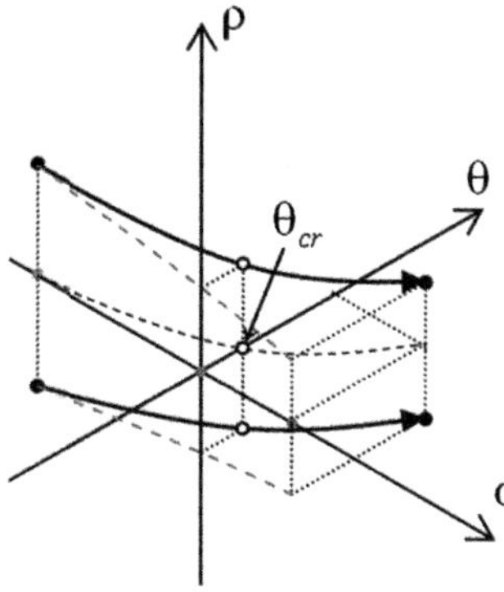

Fig. 3.1 Flutter instability – a pair of complex-conjugate eigenvalues cross the imaginary axis as the system parameter θ pass its critical value

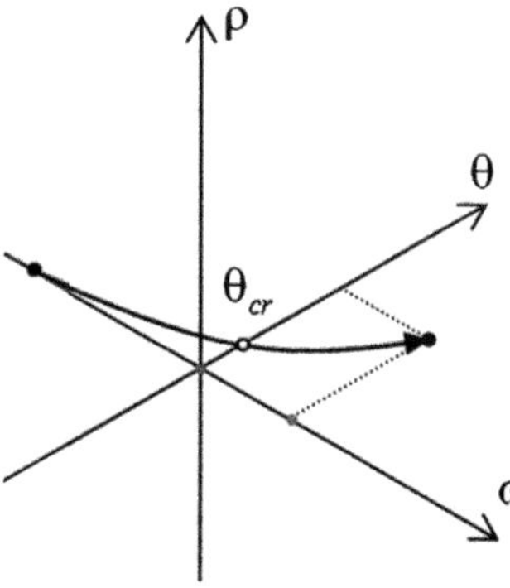

Fig. 3.2 Divergence instability – a real eigenvalue crossed the imaginary axis as the system parameter θ pass its critical value

In Sect. 2.2, we encountered situations where (as a result of frictional constraints), the inertia matrix could become singular or negative definite (i.e., when $A \leq 0$ in (3.19) or when $\det(\tilde{\mathbf{M}}) \leq 0$ in (3.24)). As mentioned above, these situations are known as Painlevé's paradoxes. We will study Painlevé's paradoxes in detail in Sect. 3.3.

For the sake of completeness, we consider here the linearized system equation for such cases. Assume that the onset of the Painlevé's paradox is at $\boldsymbol{\theta} = \boldsymbol{\theta}_{\mathrm{cr}}$ (i.e., $A(\boldsymbol{\theta}_{\mathrm{cr}}) = 0$ or $\det\left[\tilde{\mathbf{M}}(\boldsymbol{\theta}_{\mathrm{cr}})\right] = 0$). As the system parameters are varied such that $\boldsymbol{\theta}$ crosses the surface $\boldsymbol{\theta} = \boldsymbol{\theta}_{\mathrm{cr}}$ an eigenvalue goes to infinity and becomes positive as shown in Fig. 3.3. Beyond the critical value of the parameters, the solution of the linear differential equation (3.8) diverges.

3.5 Undamped Systems

In the absence of the velocity-dependent forces, i.e., $\Lambda_v = 0$, the linearized system equation (3.8) reduces to

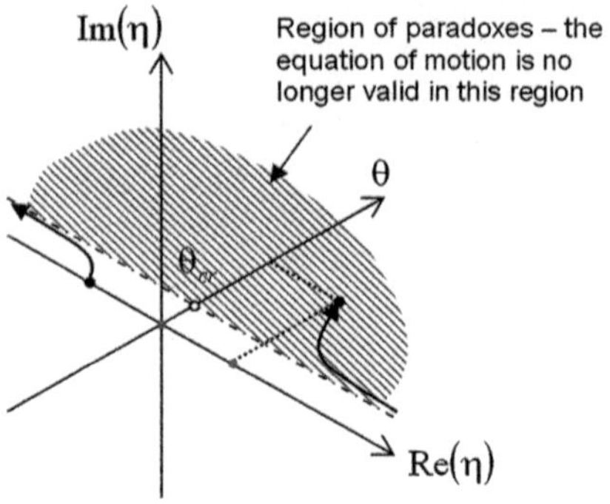

Fig. 3.3 Divergence instability due to kinematic constraint

$$\mathbf{M}\ddot{\mathbf{x}} + \Lambda_p\mathbf{x} = 0. \tag{3.27}$$

In this case, it is more convenient to assume the general solution as $\mathbf{x}(t) = \mathbf{a}e^{i\omega t}$ where $\mathbf{a}$ is a complex eigenvector. Substituting this solution into (3.27), yields

$$\left(-\omega^2\mathbf{M} + \Lambda_p\right)\cdot\mathbf{a} = 0. \tag{3.28}$$

Similar to the above discussions, non-trivial solutions are only possible when the matrix $-\omega^2\mathbf{M} + \Lambda_p$ is singular. Setting the determinant of this matrix to zero gives the characteristic equation of the undamped system:

$$\Delta\left(\omega^2\right) = \det\left(\Lambda_p - \omega^2\mathbf{M}\right) = 0. \tag{3.29}$$

The characteristic equation given by (3.29) is a polynomial of degree n in ω^2. The conditions for the divergence instability and instability due to the occurrence of the Painlevé's paradox are the same as before; i.e., $\det\left[\Lambda_p(\boldsymbol{\theta}_{\mathrm{cr}})\right] = 0$ and $\det\left[\mathbf{M}(\boldsymbol{\theta}_{\mathrm{cr}})\right] = 0$, respectively. As long as the squared natural frequencies (i.e., ${\omega_i}^2$, $i = 1\cdots n$) are real positive numbers, the origin is stable. At the onset of the flutter stability, two natural frequencies coincide and beyond the critical value of the parameters become complex conjugate (see Fig. 3.4). The condition for the flutter instability (i.e., coincidence of two squared natural frequencies) can be written as [48]

$$\frac{\partial\Delta}{\partial\omega^2} = 0. \tag{3.30}$$

Solving (3.30) together with (3.29) gives the parametric conditions (if any) for the flutter instability.

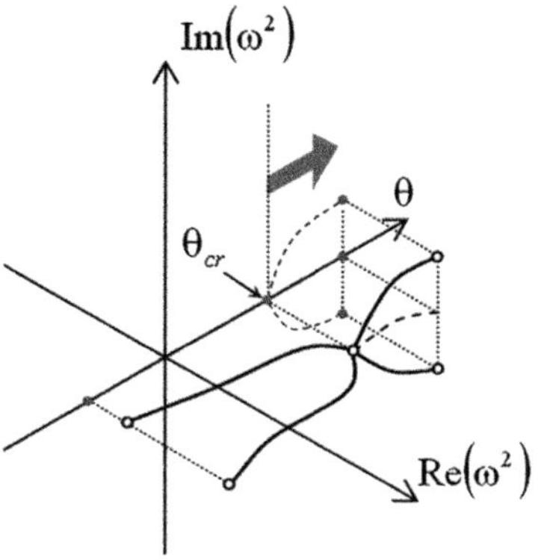

Fig. 3.4 Flutter, undamped system: coalescence of two natural frequencies

3.6 The Averaging Method

The eigenvalue analysis does not reveal much information regarding the behavior of the nonlinear system once instability occurs. The existence of periodic solutions (limit cycles), region of attraction of the stable trivial solution, and the effects of system parameters on these features as well as the size of the limit cycles (amplitude of steady-state vibrations) are important problems that cannot be solved using the linearized system's equations.

Wherever applicable,[2] we use the method of averaging [55, 56] to study the behavior of the equations of motion as a *weakly nonlinear system*. There are a few variations on the basic theorem for the first-order periodic averaging [56–58]. A slightly modified version of the theorem proven in [57] – suitable for our subsequent analyses – is presented below which establishes the error estimate of the solution of the averaged system, with respect to the that of the original differential equation.

Theorem 3.1. *(First-Order Averaging)* Consider the following system in standard form

$$\dot{\mathbf{x}} = \varepsilon \mathbf{f}(t, \mathbf{x}, \varepsilon), \quad \mathbf{x}(0) = \mathbf{x}_0. \tag{3.31}$$

Suppose

- The function $\mathbf{f} : R_+ \times D \times [0, \varepsilon_0] \to R^n$ is T-periodic with respect to t for $x \in D$. $D \subset R^n$ is an open bounded set containing the and $\varepsilon_0 > 0$ is some number.
- There exists a constant $M > 0$ such that $\|\mathbf{f}(t, \mathbf{x}, \varepsilon)\| \leq M$.
- $\mathbf{f}(t, \mathbf{x}, \varepsilon)$ is Lipschitz continuous with respect to $\mathbf{x}$ and ε with Lipschitz constants λ_x and λ_ε, respectively.

[2]The use of averaging method in this monograph is limited to the study of negative damping instability method in Sect. 4.1 and Chap. 6.

- The average, $\bar{f}(\mathbf{x}) = 1/\mathrm{T}\int_0^{\mathrm{T}} \mathbf{f}(\tau, \mathbf{x}, 0)d\tau$ exists uniformly with respect to $\mathbf{x}$.
- Consider the averaged system;

$$\dot{\mathbf{z}} = \varepsilon \bar{f}(\mathbf{z}), \quad \mathbf{z}(0) = \mathbf{x}_0. \tag{3.32}$$

The solution of (3.32), $\mathbf{z}(t; 0, \mathbf{x}_0)$, belongs to interior subset of D on time scale $1/\varepsilon$.

Then, there exists $c > 0$, $\varepsilon_0 > 0$, and $L > 0$, such that the following holds for the solutions of (3.31) and (3.32):

$$\|\mathbf{x}(t, \varepsilon) - \mathbf{z}(t, \varepsilon)\| \leq c\varepsilon.$$

For $0 \leq \varepsilon \leq \varepsilon_0$ and $0 \leq t \leq L/\varepsilon$. Also, c is independent of ε.
Proof *See Appendix A.* ■

Example 3.1 Consider the van der Pol *equation*

$$\ddot{x} + x = \varepsilon\left(1 - x^2\right)\dot{x}, \tag{3.33}$$

where $x \in R$ and $\varepsilon > 0$ is a small parameter. To convert this second-order differential equation into the standard form, (3.31), the following change of variables is used:

$$\begin{aligned} x &= r\cos(t + \beta) \\ \dot{x} &= -r\sin(t + \beta). \end{aligned} \tag{3.34}$$

Substituting (3.34) into (3.33) yields

$$\begin{aligned} \dot{r} &= \varepsilon\left(r\sin^2(t + \beta) - r^3\cos^2(t + \beta)\sin^2(t + \beta)\right), \\ \dot{\beta} &= \varepsilon\left(\sin(t + \beta)\cos(t + \beta) - r^2\cos^3(t + \beta)\sin(t + \beta)\right). \end{aligned} \tag{3.35}$$

Applying averaging to (3.35) yields

$$\begin{aligned} \dot{\bar{r}} &= \varepsilon\frac{\bar{r}}{2}\left(1 - \frac{\bar{r}^2}{4}\right). \\ \dot{\bar{\beta}} &= 0 \end{aligned} \tag{3.36}$$

Setting $\dot{\bar{r}} = 0$, the amplitude equation has two equilibrium points $\bar{r} = 0$ (trivial solution) and $\bar{r} = 2$ (periodic solution). The stability of these solutions can be established by analyzing the eigenvalues of the Jacobian of the amplitude equation.

We find that the trivial solution is *unstable* whereas the periodic solution (*i.e.*, the limit cycle) is exponentially *stable*.

In this simple case, the solution of the averaged equations, (3.36), can be written in closed-form [56], we have

$$x(t) = \bar{r}(t)\cos(t + \beta_0) + O(\varepsilon),$$

on the time scale $1/\varepsilon$ where

$$\bar{r}(t) = \frac{r_0 \mathrm{e}^{(1/2)\varepsilon t}}{\left(1 + (1/4)r_0{}^2(\mathrm{e}^{\varepsilon t} - 1)\right)^{1/2}},$$

and r_0 and β_0 are determined from the initial conditions.

The following theorem extends the $1/\varepsilon$ time scale of the validity of the $O(\varepsilon)$ accuracy of the solution of the averaged equations to infinite time scale for cases where there is attraction.

Theorem 3.2. *[57] Consider the system* (3.31*) and the averaged system (3.32). Suppose that all of the conditions of the Theorem 2.1 are satisfied. Assume further that:*

- *The averaged system has an exponentially stable equilibrium point* $\mathbf{z} = 0$.
- *The function* $\bar{f}(\mathbf{z})$ *is continuously differentiable with respect to* $\mathbf{z}$ *in* D.
- *The stable equilibrium point* $\mathbf{z} = 0$ *has a domain of attraction* $D_0 \subset D$.

If $\mathbf{x}_0 \in D_0$, *then*

$$\|\mathbf{x}(t,\varepsilon) - \mathbf{z}(t,\varepsilon)\| \le c\varepsilon,$$

for some $c>0$ *independent of* ε *and* $0 \le t < \infty$.

Proof: The proof is given in [57, Appendix II]. ■

Example 3.2: Consider again the van der Pol equation, (3.33). Alternative to the change of variable used in example 2.1, here we use the following change of variables:

$$x = r\cos\varphi,\ \dot{x} = -r\sin\varphi. \tag{3.37}$$

Substituting (3.37) into (3.33) yields

$$\begin{aligned} \dot{r} &= \varepsilon\left(r\sin^2\varphi - r^3\cos^2\varphi\sin^2\varphi\right) \\ \dot{\varphi} &= 1 + \varepsilon\left(\sin\varphi\cos\varphi - r^2\cos^3\varphi\sin\varphi\right). \end{aligned} \tag{3.38}$$

Assuming $\dot{\varphi} \neq 0$ and dividing the two equations in (3.38) gives

$$\frac{\mathrm{d}r}{\mathrm{d}\varphi} = \varepsilon\left(\frac{r\sin^2\varphi - r^3\cos^2\varphi\sin^2\varphi}{1+\varepsilon(\sin\varphi\cos\varphi - r^2\cos^3\varphi\sin\varphi)}\right),$$

which is in the standard form (3.31) with φ as the time-like parameter. The averaged equation is found as

$$\frac{\mathrm{d}\bar{r}}{\mathrm{d}\varphi} = \varepsilon\frac{\bar{r}}{2}\left(1 - \frac{\bar{r}^2}{4}\right). \tag{3.39}$$

We have already seen that $\bar{r} = 2$ is an exponentially attractive equilibrium point of (3.39). Based on Theorem 3.2 we have $r(\varphi) - \bar{r}(\varphi) = O(\varepsilon)$ for $\varphi \in (0, \infty)$.

Chapter 4
Friction-Induced Instability

In this chapter, we take a close look at the three distinct friction-induced instability mechanisms mentioned in Chap. 1. We start with the negative damping instability mechanism in Sect. 4.1. To demonstrate the role of friction, we analyze the well-known mass-on-a-conveyor model with a decreasing coefficient of friction function with sliding velocity. Both eigenvalue analysis method and the method of averaging are used to study this system. The material covered in this section serves as an introduction to Chap. 5 where we study the negative damping instability mechanism in the lead screw drives.

In Sect. 4.2, the mode coupling instability mechanism is considered. In Chap. 3, we have seen the effect of nonconservative forces in creating circulatory systems capable of exhibiting flutter instability. Examples are presented in this section to study the flutter instability with or without friction. Material presented in this section is a prelude to Chap. 7 where we study the mode coupling instability in the lead screw drives.

Finally, in Sect. 4.3, we turn our attention to the kinematic constraint instability. This section begins by the introduction of the Painlevé's paradoxes which play an important role in the kinematic constraint instability mechanism. The self-locking property which is another consequence of friction in the rigid body dynamics is also discussed in this section. This effect has a prominent presence in the study of the lead screws in Chaps. 7 and 8. The concepts presented here form the basis for the study of the kinematic constraint instability in the lead screws in Chap. 8.

4.1 Negative Damping Instability

The negative slope in the friction–sliding velocity curve or the difference between static and kinematic coefficients of friction can lead to the so-called stick–slip vibrations (see, e.g., [14, 59]). In most instances, researchers adopted the well-known mass-on-a-conveyor model to study the stick–slip vibrations (see, e.g., [17, 60, 61]). In this section, we will also consider this simple model – as shown in Fig. 4.1 – to investigate the effects of the negative damping instability mechanism.

O. Vahid-Araghi and F. Golnaraghi, *Friction-Induced Vibration in Lead Screw Drives*,
DOI 10.1007/978-1-4419-1752-2_4,

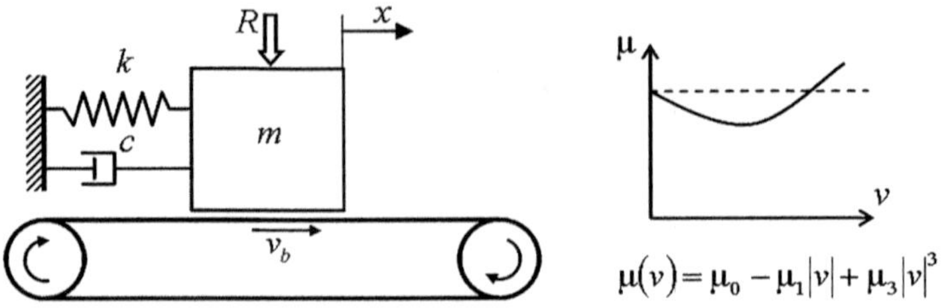

Fig. 4.1 1-DOF mass-on-a-conveyor model

As shown in Fig. 4.1, a block of mass m is held by a linear spring k and a linear damper c. The block slides against a moving conveyor that has a constant velocity $v_b > 0$. Here, following [57, 62], the coefficient of friction is assumed to be a cubic function of relative velocity:

$$\mu(v) = \mu_0 - \mu_1|v| + \mu_3|v|^3, \quad \mu_0, \mu_1, \mu_3 > 0, \tag{4.1}$$

where $v = v_b - \dot{x}$ is the relative sliding velocity. The equation of motion for this model can be written as [62]

$$\text{If } \dot{x} \neq v_b \quad \text{then} \quad m\ddot{x} + c\dot{x} + kx = N\mu(v_b - \dot{x})\text{sgn}(v_b - \dot{x}) \text{ (slip)}, \tag{4.2}$$

$$\text{If } \dot{x} = v_b \quad \text{then} \quad \ddot{x} = 0 \quad \text{and} \quad c\dot{x} + kx < N\mu_0 \text{ (stick)}, \tag{4.3}$$

where $N > 0$ is the normal force between the mass and the conveyor. Transferring the steady-sliding state to the origin yields

$$m\ddot{y} + c\dot{y} + ky = N[\mu(v_b - \dot{y})\text{sgn}(v_b - \dot{y}) - \mu(v_b)], \tag{4.4}$$

where $y = x - x_0$ and $x_0 = (N/k)\mu(v_b)$.

Considering small perturbations around the steady-sliding equilibrium point where $v_b - \dot{y} > 0$, the linearized equation of motion is found from (4.4) as

$$m\ddot{y} + (c + d_\mu N)\dot{y} + ky = 0, \tag{4.5}$$

where

$$d_\mu = \left.\frac{\mathrm{d}\mu(v)}{\mathrm{d}v}\right|_{v=v_b>0} = -\mu_1 + 3v_b^2\mu_3 \tag{4.6}$$

is the slope of the coefficient of friction vs. relative velocity. The eigenvalues corresponding to the linear differential equation (4.5) are found as

$$\lambda_{1,2} = \frac{-(c + d_\mu N) \pm \sqrt{(c + d_\mu N)^2 - 4mk}}{2m}.$$

It is obvious that when d_μ is negative and $c < -Nd_\mu$, the origin of system (4.5) is unstable. In this situation, the vibration amplitude grows until it reaches an *attractive* limit cycle. If trajectories reach the stick boundary, i.e., $v_b - \dot{y} = 0$, stick–slip periodic vibration occurs. In the next two sections, periodic vibrations in cases where pure-slip and stick–slip motions occur are studied separately. In these sections, a perturbation method (i.e., the method of averaging) is used to construct asymptotic solutions since due to nonlinearity and discontinuity of (4.4), closed-form solutions are not available.

4.1.1 Periodic Vibration: Pure-Slip Motion

The following derivations closely follow [57, 62]. If a periodic motion exists where $v > 0$ for all $t > 0$, then no sticking occurs and the discontinuity of the friction is not encountered. In this case, (4.4) simplifies to

$$\begin{aligned} m\ddot{y} + c\dot{y} + ky &= N\Big(\mu(v_b - \dot{y}) - \mu(v_b)\Big), \\ &= N\Big(\mu_1\dot{y} + \mu_3(v_b - \dot{y})^3 - \mu_3 v_b^3\Big). \end{aligned} \tag{4.7}$$

The method of first-order averaging applies to *weakly nonlinear* systems.[1] This requirement is fulfilled here since friction and damping coefficients are small quantities. The first step is to convert (4.7) to a nondimensional form. Let

$$\omega = \sqrt{\frac{k}{m}}, \quad \tau = \omega t, \quad \tilde{c} = \frac{c}{m\omega}, \quad L = \frac{N}{k}, \quad y = Lu. \tag{4.8}$$

Substituting these relationships into (4.7) yields

$$u'' + u = \left(-\tilde{c} + L\omega\mu_1 - 3L\omega\mu_3 v_b^2\right)u' + 3L^2\omega^2\mu_3 v_b u'^2 - L^3\omega^3\mu_3 u'^3. \tag{4.9}$$

To explicitly identify "size" of the system parameters, a formal book-keeping parameter $0 < \varepsilon \ll 1$ is introduced. Let

$$\varepsilon\hat{c} = \tilde{c}, \quad \varepsilon\hat{\mu}_1 = \mu_1, \quad \varepsilon\hat{\mu}_3 = \mu_3, \tag{4.10}$$

where $\hat{c}$, $\hat{\mu}_1$, $\hat{\mu}_3$, and all other parameters are $O(1)$ with respect to ε. Substituting (4.10) into (4.9) yields

$$u'' + u = \varepsilon h(u'), \tag{4.11}$$

[1] Refer to Sect. 3.6.

where

$$h(u') = \left(-\widehat{c} + \widehat{\kappa_1}\right)u' + \widehat{\kappa}_2 u'^2 - \widehat{\kappa}_3 u'^3, \tag{4.12}$$

where

$$\widehat{\kappa}_1 = L\omega\widehat{\mu}_1 - 3L\omega\widehat{\mu}_3 {v_b}^2, \quad \widehat{\kappa}_2 = 3L^2\omega^2\widehat{\mu}_3 v_b, \quad \widehat{\kappa}_3 = L^3\omega^3\widehat{\mu}_3. \tag{4.13}$$

The damping and nonlinear terms appear in (4.11) as *small* perturbation to a harmonic oscillator. In order to apply the method of averaging, (4.11) must be converted into the standard form.[2] To accomplish this, the following change of variables is used:

$$u = a\cos(\tau + \varphi), \; u' = -a\sin(\tau + \varphi). \tag{4.14}$$

Substituting (4.14) into (4.11) yields

$$a' = -\varepsilon h\left(-a \sin(\tau + \varphi)\right) \sin(\tau + \varphi), \tag{4.15}$$

$$a\varphi' = -\varepsilon h\left(-a \sin(\tau + \varphi)\right) \cos(\tau + \varphi). \tag{4.16}$$

The first-order averaged equations are obtained by averaging the right-hand side of (4.15) and (4.16) over a period (i.e., $T = 2\pi$) while keeping a and ϕ constant:

$$\begin{aligned} a' &= -\frac{\varepsilon}{2\pi}\int_0^{2\pi} h\left(-a\sin(\tau + \varphi)\right) \sin(\tau + \varphi)\mathrm{d}\tau, \\ \varphi' &= -\frac{\varepsilon}{2\pi a}\int_0^{2\pi} h\left(-a\sin(\tau + \varphi)\right) \cos(\tau + \varphi)\mathrm{d}\tau. \end{aligned} \tag{4.17}$$

Carrying out the integrations yields

$$\begin{aligned} a' &= \frac{\varepsilon}{2}\left(-\widehat{c} + \widehat{\kappa}_1\right)a - \frac{3\varepsilon}{8}\widehat{\kappa_3}a^3, \\ \varphi' &= 0. \end{aligned} \tag{4.18}$$

The averaged amplitude equations have two equilibrium points: A trivial solution, that is, $a = 0$, corresponding to the steady-sliding equilibrium; and a nontrivial solution (i.e., a limit cycle) is given by

$$a = a_1 = 2\sqrt{\frac{-\widehat{c} + \widehat{\kappa}_1}{3\widehat{\kappa}_3}}, \quad \widehat{c} - \widehat{\kappa}_1 < 0. \tag{4.19}$$

---

[2]See Theorem 3.1.

The trivial solution is *asymptotically* stable if $\mathrm{d}a'/\mathrm{d}a|_{a=0} < 0$. From (4.18), one may find

$$\begin{aligned}\left.\frac{\mathrm{d}a'}{\mathrm{d}a}\right|_{a=0} &= \frac{\varepsilon}{2}\left(-\hat{c} + \hat{\kappa}_1\right),\\ &= \frac{1}{2m\omega}\left(-c + N\mu_1 - 3N\mu_3 v_{\mathrm{b}}^2\right),\\ &= \frac{1}{2m\omega}\left(-c + Nd_\mu\right).\end{aligned} \tag{4.20}$$

where d_μ is given by (4.6). From (4.20), the condition for the stability of the trivial solution is found to be $-c + Nd_\mu < 0$ which agrees with what was found from linear eigenvalue analysis of the previous section. The limiting value of the belt constant velocity from inequality (4.20) is found as

$$v_{\mathrm{b\_max}} = \sqrt{\frac{\mu_1 N - c}{3\mu_3 N}}, \quad \mu_1 N - c > 0.$$

For $v_{\mathrm{b}} > v_{\mathrm{b\_max}}$, the steady-sliding state (i.e., trivial solution) is stable. Note that if $c > \mu_1 N$ then the trivial solution is stable for all values of the belt velocity.

If the trivial equilibrium point is unstable, i.e., $-\hat{c} + \hat{\kappa}_1 > 0$, according to (4.19) there is nontrivial solution; $a = a_1$ (limit cycle in the original system's phase plane). The stability of this solution is assessed by evaluating

$$\begin{aligned}\left.\frac{\mathrm{d}a'}{\mathrm{d}a}\right|_{a=a_1} &= \frac{\varepsilon}{2}\left(-\hat{c} + \hat{\kappa}_1\right) - \frac{9\varepsilon}{8}\hat{\kappa}_3 a_1^2,\\ &= -\varepsilon\left(-\hat{c} + \hat{\kappa}_1\right) < 0.\end{aligned}$$

Thus, when the origin is *unstable*, trajectories are attracted by a *stable* limit cycle. There is, however, one more step needed before accepting the above nontrivial pure-slip solution: The condition $v > 0$ (i.e., pure-slip motion) must be checked. In terms of the nondimensional system parameters, this condition is satisfied when $L\omega \max(u') < v_{\mathrm{b}}$. The maximum velocity of the mass according to the first-order averaged solution is $\max(u') = a_1$, thus from (4.19) we must have

$$2\sqrt{\frac{-\hat{c} + \hat{\kappa}_1}{3\hat{\kappa}_3}} < \frac{v_{\mathrm{b}}}{L\omega}.$$

Using (4.8) and (4.10), the lower limit of belt velocity for the existence of pure-slip periodic vibrations is found as

$$v_{\mathrm{b\_min}} = \sqrt{\frac{4}{5}\frac{\mu_1 N - c}{3\mu_3 N}} = \sqrt{\frac{4}{5}}v_{\mathrm{b\_max}}.$$

Based on the above findings, the pure-slip periodic vibration can only occur if the belt velocity is within certain limits; $v_{b\_min} < v_b < v_{b\_max}$. For lower belt velocities, i.e., $v_b < v_{b\_min}$, stick–slip vibration occurs which is characterized by periodic sticking of the mass to the conveyor belt. This case is considered in the next section.

4.1.2 *Periodic Vibration: Stick–Slip Motion*

In the stick–slip vibrations, the mass moves with the belt for a part of one period. This intermittent motion results in a dynamical system with *varying* degrees-of-freedom (DOFs). Obviously, during the stick phase, the number of DOFs is zero and it is one otherwise [see (4.2) and (4.3)]. Due to this discontinuity, the averaging method used in previous section is no longer applicable. A common but intricate approach to construct an approximate solution for the stick–slip motion is to treat the stick phase and slip phase separately and then *stitch* the two results together [57, 62]. Here, however, we take a different approximation approach; *smoothing*.[3]

To avoid the discontinuity of the friction with respect to the relative velocity, the coefficient of friction – given by (4.1) – is modified to

$$\mu(v) = \mu_0\left(1 - e^{-r|v|}\right) - \mu_1|v| + \mu_3|v|^3. \tag{4.21}$$

As shown in Fig. 4.2, the introduction of the second bracket brings the coefficient of friction rapidly to zero in a small range of relative velocities. The parameter (r) controls the steepness of the coefficient of friction function in this small range.

This smoothing modification to the coefficient of friction function enables us to apply the method of first-order averaging to the cases where relative velocity becomes zero or even changes sign. There are, however, side effects that must be treated with caution:

Remark 4.1. The *stick–slip* vibration is now replaced by a quasi-stick–slip vibration. Because of smoothing, no sticking occurs; however, the relative velocity remains *close* to zero in a portion of a period resembling that of a stick–slip cycle. Parameter r directly affects how close quasi-stick phase gets to a true stick phase. □

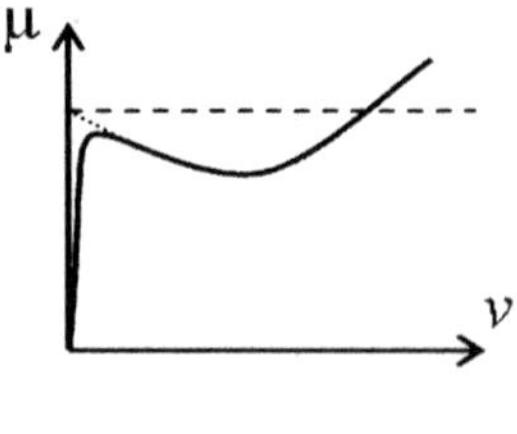

Fig. 4.2 Smoothed coefficient of friction function

[3]Smoothed friction models were used by others, see, for example, [61, 63–67].

Remark 4.2. For small belt velocities, in a small neighborhood of zero relative velocity, the origin (i.e., steady-sliding state) is *stable* regardless of the value of other system parameters. Because of the smoothing, a boundary layer is created where the gradient of the coefficient of friction function is always positive; $\mathrm{d}\mu/\mathrm{d}v|_{v=0} = r\mu_0 - \mu_1 > 0$. Obviously, the current approach is not suitable for the study of the behavior of the system inside this boundary layer. □

Remark 4.3. The parameter r can be very large thus diminishing the boundary layer effects. This parameter, however, cannot be larger than $O(1)$ in order to insure $O(\varepsilon)$ accuracy of the first-order averaging results. □

Substituting (4.21) into (4.4) and using (4.8) yields

$$\begin{aligned} u'' + u = & -\tilde{c}u' + \mu_0\left(1 - \mathrm{e}^{-\hat{r}|1-\rho u'|}\right)\mathrm{sgn}(1-\rho u') - \mu_1 v_\mathrm{b}(1-\rho u') \\ & + \mu_3 v_\mathrm{b}^3(1-\rho u')^3 - \mu_\mathrm{ss}, \end{aligned} \tag{4.22}$$

where

$$\rho = \frac{L\omega}{v_\mathrm{b}}, \qquad \hat{r} = r v_\mathrm{b}$$

and the abbreviation $\mu_\mathrm{ss} = \mu(v_\mathrm{b})$ was used. Similar to the previous section, the small parameter $\varepsilon \ll 1$ is introduced into (4.22) through the use of (4.10) also, the following new parameters are defined:

$$\varepsilon\hat{\mu}_0 = \mu_0, \qquad \varepsilon\hat{\mu}_\mathrm{ss} = \mu_\mathrm{ss}, \tag{4.23}$$

$\hat{c}$, $\hat{\mu}_0$, $\hat{\mu}_1$, $\hat{\mu}_3$, $\hat{\mu}_\mathrm{ss}$, and all other parameters are assumed to be $O(1)$ with respect to ε. Substituting (4.10) and (4.23) into (4.22) and rearranging gives

$$u'' + u = \varepsilon h(u'),$$

where

$$\begin{aligned} h(u') = & -\hat{c}u' + \hat{\mu}_0\left(1 - \mathrm{e}^{-\hat{r}|1-\rho u'|}\right)\mathrm{sgn}(1-\rho u') - \hat{\mu}_1 v_\mathrm{b}(1-\rho u') \\ & + \hat{\mu}_3 v_\mathrm{b}^3(1-\rho u')^3 - \hat{\mu}_\mathrm{ss}. \end{aligned} \tag{4.24}$$

Compared with (4.12), (4.24) is considerably more complicated. Limiting our studies to the case where the moving mass does not overtake the belt, i.e., $0 \leq \rho u' \leq 1$, (4.24) is simplified to

$$h(u') = \left(-\hat{c} + \hat{\kappa}_1\right)u' + \hat{\kappa}_2 u'^2 - \hat{\kappa}_3 u'^3 + \hat{\kappa}_4\left(1 - \mathrm{e}^{\hat{r}\rho u'}\right), \tag{4.25}$$

where

$$\widehat{\kappa}_1 = \rho v_b \widehat{\mu}_1 - 3\rho \widehat{\mu}_3 v_b^3, \quad \widehat{\kappa}_2 = 3\rho^2 \widehat{\mu}_3 v_b^3, \quad \widehat{\kappa}_3 = \rho^3 v_b^3 \widehat{\mu}_3, \quad \text{and}$$
$$\widehat{\kappa}_4 = \widehat{\mu}_0 \mathrm{e}^{-\widehat{r}}. \tag{4.26}$$

Note that the first three terms of (4.25) are identical to (4.12). Substituting (4.25) into (4.17) and carrying out the integration, the first-order averaged amplitude equation is found as

$$a' = \varepsilon f(a) = \frac{\varepsilon}{2}\left(-\widehat{c} + \widehat{\kappa}_1\right)a - \frac{3\varepsilon}{8}\widehat{\kappa}_3 a^3 - \varepsilon\widehat{\kappa}_4 \Lambda\left(\widehat{r}\rho a\right), \tag{4.27}$$

where

$$\Lambda(\xi) = \frac{1}{2\pi}\int_0^{2\pi} \sin\psi \, \mathrm{e}^{\xi \sin\psi} \mathrm{d}\psi = \sum_{n=1}^{\infty} \frac{n}{2^{2n-1}(n!)^2}\xi^{2n-1}.$$

The averaged amplitude equation given by (4.27) is considerably more complex than what we have found in the previous section, i.e., (4.18), when discontinuity was not encountered. It is interesting to note that if $a\rho < 1$, then $\mathrm{e}^{-\widehat{r}}\Lambda\left(\widehat{r}\rho a\right) = 1/2\pi \int_0^{2\pi} \sin\psi \, \mathrm{e}^{\widehat{r}(a\rho \sin\psi - 1)} \mathrm{d}\psi \downarrow 0$ as $\widehat{r} \to \infty$. As a result, (4.27) simplifies to (4.18), in the limit.

As expected, $a = 0$ is the trivial solution of (4.27). Similar to the case of the pure-slip motion, the stability of the steady-sliding equilibrium point (i.e., the origin) is evaluated from the sign of $\mathrm{d}a'/\mathrm{d}a|_{a=0}$. From (4.27) one finds

$$\begin{aligned}\left.\frac{\mathrm{d}a'}{\mathrm{d}a}\right|_{a=0} &= \frac{\varepsilon}{2}\left(-\widehat{c} + \widehat{\kappa}_1 - \widehat{\kappa}_4 \widehat{r}\rho\right), \\ &= \frac{1}{2m\omega}\left(-c + N\mu_1 - 3N\mu_3 v_b^2 - rN\mu_0 \mathrm{e}^{-rv_b}\right), \\ &= \frac{1}{2m\omega}\left(-c + Nd_\mu\right).\end{aligned} \tag{4.28}$$

where, instead of (4.6), d_μ is given by

$$d_\mu = \left.\frac{\mathrm{d}\mu(v)}{\mathrm{d}v}\right|_{v=v_b>0} = -r\mu_0 \mathrm{e}^{-rv_b} + \mu_1 - 3\mu_3 v_b^2.$$

Although the smoothing of the coefficient of friction function has modified the slope of this curve, (4.28) is similar to (4.20).

Due to the complexity of (4.27), nontrivial solutions – if any – may only be found numerically. However, the existence and the number of nontrivial solutions can be determined by examining (4.27) more closely. Assuming $a \neq 0$, dividing (4.27) by a, and expanding gives

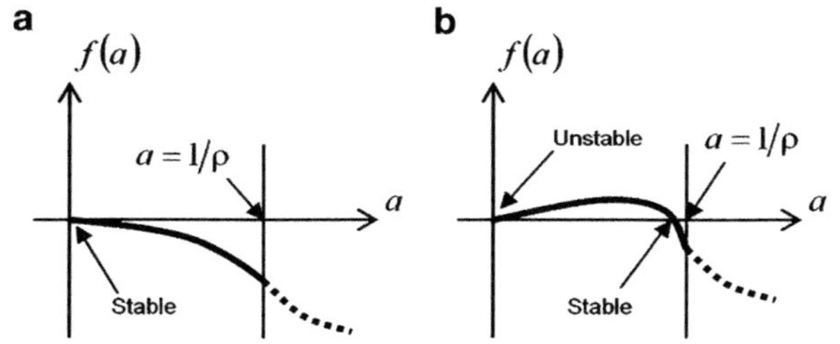

Fig. 4.3 Variation of the averaged amplitude equation

$$\frac{a'}{a} = \varepsilon\left(b_0 + b_2a^2 + \sum_{n=1}^{\infty} b_{2n+2}a^{2n+2}\right), \quad a > 0, \tag{4.29}$$

where

$$\begin{aligned} b_0 &= \frac{1}{2}\left(-\widehat{c} + \widehat{\kappa}_1 - \widehat{r}\rho\widehat{\kappa}_4\right), \\ b_2 &= -\frac{1}{16}\left(6\widehat{\kappa}_3 + \widehat{r}^3\rho^3\widehat{\kappa}_4\right), \\ b_{2n+2} &= -\widehat{\kappa}_4\frac{(n+2)\,\widehat{r}^{2n+2}\rho^{2n+2}}{2^{2n+2}((n+2)!)^2}a^{2n-2}, \quad n \geq 1, \end{aligned} \tag{4.30}$$

The nontrivial solutions are found by setting (4.29) to zero;

$$b_0 + b_2\left(a^2\right) + \sum_{n=1}^{\infty} b_{2n+2}\left(a^2\right)^{n+1} = 0, \tag{4.31}$$

which is a polynomial equation in a^2. Note that from (4.26) and (4.30) and the initial assumption on the system parameters (i.e., parameters are assumed to be positive), it is found that $b_2 < 0$ and $b_{2n+2} < 0, \quad n = 1, 2, \ldots$. Consequently, if $b_0 < 0$ (i.e., stable trivial solution), then there are no sign changes in the coefficients of polynomial (4.31). From Descartes' rule of signs [68], it is deduced that (4.31) has no positive solutions and hence the averaged system has no nontrivial equilibrium points. On the other hand, if $b_0 > 0$, then there is one sign change in the coefficients of polynomial (4.31) resulting in one positive solution and one nontrivial equilibrium point which of course corresponds to a periodic solution of the original system. Schematic plots of (4.27) for these two cases are shown in Fig. 4.3. In Fig. 4.3a, the trivial solution is stable while in Fig. 4.3b the trivial solution is unstable and there is a *stable* nontrivial solution (i.e., an attractive limit cycle).

Remark 4.4. For $a > 1/\rho$, the simplifying assumption (i.e., $u' < v_b/\omega L$ or $\dot{y} < v_b$) leading to (4.25) is violated. This is the case where the moving mass overtakes the conveyor for a portion of a period. To obtain the first-order averaged equation, the averaging must be applied directly to (4.24). The averaged amplitude of vibrations can be found from the resulting equation, numerically. □

A numerical example is given next to demonstrate the utility of the averaging method described above.

4.1.3 A Numerical Example

The equation of motion (4.4) describes the system shown in Fig. 4.1. In this section, results of numerical simulations with both discontinuous and smoothed (continuous) coefficient of friction functions are presented and compared with the averaging results. The numerical values of the parameters are given in Table 4.1.

Figure 4.4 shows the variation of the discontinuous coefficient of friction, (4.1), and the smoothed coefficient of friction function, (4.21), for two values of the parameter r. As can be seen, the approximation of the smoothed function is considerably improved by increasing the value of r from 20 to 200. The stick–slip limit cycle of the discontinuous system is shown in Fig. 4.5. The stick boundary is shown in this figure by a horizontal dashed line at $u' = 1/\rho$ (i.e., $\dot{y} = v_b$). In this figure, two more limit cycles are shown which correspond to the system with

Table 4.1 Numerical value of the parameters used in the simulations

Parameter	Value	Parameter	Value
m	1 kg	μ_0	0.4
k	10 N/m	μ_1	0.45 s/m
c	0.1 N s/m	μ_3	0.6 s^3/m^3
N	1 N		

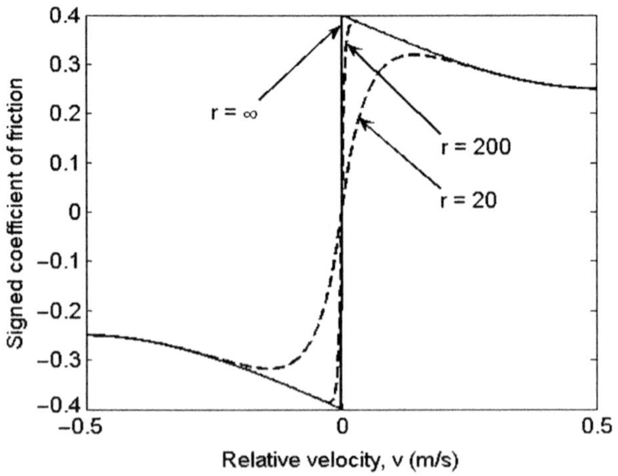

Fig. 4.4 Discontinuous and smoothed coefficient of friction functions

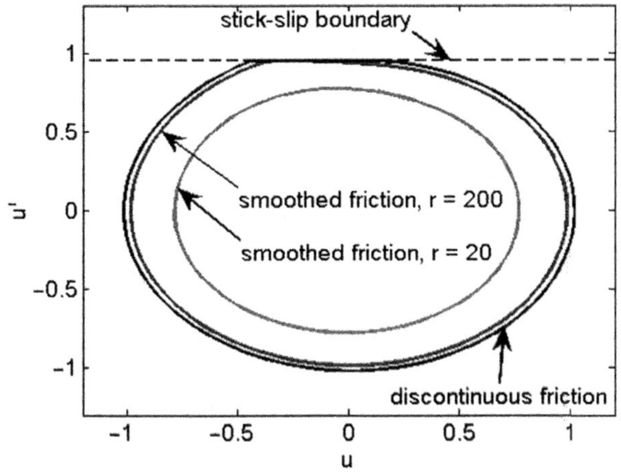

Fig. 4.5 Periodic motion of the system with discontinuous and smoothed friction function

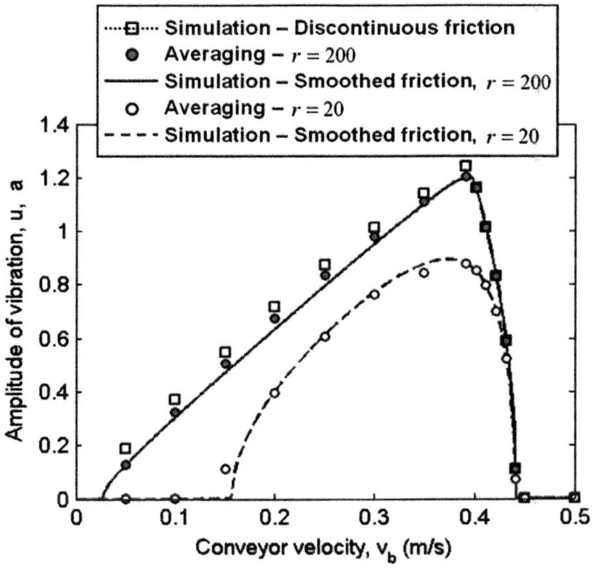

Fig. 4.6 Comparison of simulation and averaging results

smoothed coefficient of friction. As expected, for higher values of r, the system trajectory is very close to that of the discontinuous system.

Figure 4.6 shows the averaging results for the two values of the parameter r as the conveyor velocity is varied. In this figure, simulation results from the discontinuous and smoothed models are also included for comparison.

For the two cases of $r = 20$ and $r = 200$, the averaging results very accurately estimate the amplitude of vibrations when compared with the respective numerical simulation results of the model with smoothed friction function.

The difference between the results obtained from the system with smoothed coefficient of friction, and the original system (discontinuous friction) is significant for $r = 20$. This, of course, is the same as the difference shown in Fig. 4.5. For $r = 200$, on the other hand, the differences among numerical simulation of the original equations, numerical simulation of the smoothed equations, and the averaging results are much smaller.

4.1.4 Further References on Negative Damping

Using an exponentially decreasing model for the coefficient of friction, Hetzler et al. [61] used the method of averaging to study the steady-state solutions of a system similar to the one shown in Fig. 4.1. They showed that as damping is increased, the unstable steady-sliding equilibrium point goes through a subcritical Hopf bifurcation [69], resulting in an unstable limit cycle that defines the region of attraction of the stable equilibrium point.

Thomsen and Fidlin [62] also used averaging techniques to derive approximate expressions for the amplitude of stick–slip and pure-slip (when no sticking occurs) vibrations in a model similar to Fig. 4.1. They used a third-order polynomial to describe the velocity-dependent coefficient of friction. Other researchers have shown that in cases where the coefficient of friction is a nonlinear function of sliding velocity (e.g., humped friction model), the presence of one or more sections of negative slope in the friction–sliding velocity curve can lead to self-excited vibration without sticking [4, 70, 71].

4.2 Mode Coupling

In Chap. 3, we mentioned circulatory systems which are described (after linearization) by asymmetric stiffness and/or damping coefficient matrices. Stability of these class of systems has been studied by many authors (see, e.g., [48, 49, 72, 73]). In Sect. 4.2.1 below, we give a classic example where a follower force causes flutter instability. In multi-DOF systems, friction force may act as a follower force and destroy the symmetry of the stiffness and damping matrices resulting in flutter instability known as the mode coupling instability mechanism. This mechanism was first used to explain brake squeal [7]. Ono et al. [74] and Mottershead and Chan [75] studied hard disk drive instability using a similar concept. In Sect. 4.2.2, we study the mode coupling instability mechanism in a simple 2-DOF system with friction.

4.2.1 Example No. 1: Flutter Instability

Figure 4.7 shows a 2-DOF planar manipulator. A force, P, is applied to the free end in such a way that it is always aligned with the second link (i.e., a follower force). The equations of motion of this system can be written as

$$\begin{aligned}(m_1+m_2)l^2\ddot{\theta}_1+m_2l^2\ddot{\theta}_2\cos(\theta_2-\theta_1)-m_2l^2\dot{\theta}_2^2\sin(\theta_2-\theta_1)\\+k_1\theta_1-k_2(\theta_2-\theta_1-\theta_0)=-Pl\sin(\theta_2-\theta_1)\\m_2l^2\ddot{\theta}_2+m_2l^2\ddot{\theta}_1\cos(\theta_2-\theta_1)+m_2l^2\dot{\theta}_1^2\sin(\theta_2-\theta_1)+k_2(\theta_2-\theta_1-\theta_0)=0.\end{aligned} \tag{4.32}$$

In matrix form, (4.32) can be written as

$$\widehat{M}(\boldsymbol{\theta})\ddot{\boldsymbol{\theta}}+\mathbf{h}(\boldsymbol{\theta},\dot{\boldsymbol{\theta}})+\mathbf{K}\boldsymbol{\theta}=\mathbf{f}(\boldsymbol{\theta}),$$

where $\boldsymbol{\theta}=[\theta_1 \quad \theta_2]^{\mathrm{T}}$ and

$$\widehat{\mathbf{M}}(\boldsymbol{\theta})=\begin{bmatrix}(m_1+m_2)l^2 & m_2l^2\cos(\theta_2-\theta_1)\\ m_2l^2\cos(\theta_2-\theta_1) & m_2l^2\end{bmatrix},$$

$$\mathbf{h}(\boldsymbol{\theta},\dot{\boldsymbol{\theta}})=\begin{bmatrix}-m_2l^2\dot{\theta}_2^2\sin(\theta_2-\theta_1)\\+m_2l^2\dot{\theta}_1^2\sin(\theta_2-\theta_1)\end{bmatrix},$$

$$\mathbf{K}=\begin{bmatrix}k_1+k_2 & -k_2\\ -k_2 & k_2\end{bmatrix},$$

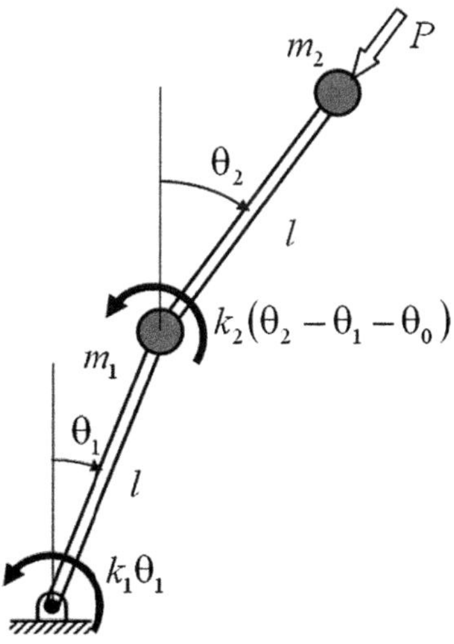

Fig. 4.7 2-DOF manipulator with a follower force

$$\mathbf{f}(\boldsymbol{\theta}) = \begin{bmatrix} -Pl\,\sin(\theta_2 - \theta_1) - k_2\theta_0 \\ k_2\theta_0 \end{bmatrix}.$$

The equilibrium configuration is found by setting all temporal derivatives to zero. We have

$$\begin{bmatrix} k_1 + k_2 & -k_2 \\ -k_2 & k_2 \end{bmatrix} \begin{bmatrix} \theta_{1\text{eq}} \\ \theta_{2\text{eq}} \end{bmatrix} = \begin{bmatrix} -Pl\,\sin\left(\theta_{2\text{eq}} - \theta_{1\text{eq}}\right) - k_2\theta_0 \\ k_2\theta_0 \end{bmatrix},$$

which yields

$$\theta_{1\text{eq}} = -\frac{Pl}{k_1}\sin\theta_0,$$

$$\theta_{2\text{eq}} = -\frac{Pl}{k_1}\sin\theta_0 + \theta_0.$$

The linearized system of equations with respect to small vibrations around the equilibrium state is found as

$$\mathbf{M}\boldsymbol{\varphi} + \boldsymbol{\Lambda}_{\text{p}}\boldsymbol{\varphi} = \mathbf{0},$$

where $\boldsymbol{\varphi} = [\phi_1 \quad \phi_2]^{\text{T}}$, $\phi_1 = \theta_1 - \theta_{1\text{eq}}$, $\phi_2 = \theta_2 - \theta_{2\text{eq}}$, and

$$\mathbf{M} = \widehat{\mathbf{M}}\left(\boldsymbol{\theta}_{\text{eq}}\right) = \begin{bmatrix} (m_1 + m_2)l^2 & m_2 l^2 \cos\theta_0 \\ m_2 l^2 \cos\theta_0 & m_2 l^2 \end{bmatrix},$$

$$\boldsymbol{\Lambda}_{\text{p}} = \mathbf{K} + \left.\frac{\partial \mathbf{f}}{\partial \boldsymbol{\theta}}\right|_{\boldsymbol{\theta}=\boldsymbol{\theta}_{\text{eq}}} = \begin{bmatrix} k_1 + k_2 - Pl\,\cos\theta_0 & -k_2 + Pl\,\cos\theta_0 \\ -k_2 & k_2 \end{bmatrix}.$$

First, we notice that the inertia matrix is symmetric and positive definite (i.e., $(m_1 + m_2)l^2 > 0$ and $\det(\mathbf{M}) = l^4 m_2\left(m_1 + m_2\sin^2\theta_0\right) > 0$). The divergence instability is ruled out since $\det(\mathbf{K}) = k_1 k_2 > 0$. To check for the possibility of flutter instability, first we need to derive the characteristic equation. According to (3.29) we have

$$\Delta\left(\omega^2\right) = a_4\omega^4 + a_2\omega^2 + a_0 = 0, \tag{4.33}$$

where

$$a_4 = \det(\mathbf{M}) = l^4 m_2\left(m_1 + m_2\sin^2\theta_0\right) > 0, \tag{4.34}$$

$$a_2 = Pl^3 m_2\cos\theta_0(1 + \cos\theta_0) - 2m_2 l^2 k_2(1 + \cos\theta_0) - l^2\left(m_1 k_2 + m_2 k_1\right), \tag{4.35}$$

$$a_0 = \det(\Lambda_p) = k_1 k_2 > 0. \tag{4.36}$$

Following (3.30), the frequency at the flutter boundary is found as

$$\frac{\partial \Delta(\omega^2)}{\partial(\omega^2)} = 2a_4\omega^2 + a_2 = 0 \rightarrow \omega^2 = -\frac{a_2}{2a_4}.$$

Substituting this result into (4.33) gives

$$\Delta(\omega^2) = -\frac{{a_2}^2}{4a_4} + a_0 = 0 \rightarrow {a_2}^2 - 4a_0a_4 = 0, \tag{4.37}$$

which is the same as requiring the discriminant of (4.33) to be zero. Solving (4.37) for P results in the critical load value for the onset of the flutter instability. Substituting (4.34)–(4.36) into (4.37) result in a quadratic equation in P;

$$(b_1P - b_2)^2 - 4b_3 = 0 \rightarrow P = \frac{b_2 \pm 2\sqrt{b_3}}{b_1},$$

where

$$\begin{aligned} b_1 &= l^3 m_2 \cos\theta_0 (1 + \cos\theta_0) > 0, \\ b_2 &= 2m_2 l^2 k_2 (1 + \cos\theta_0) + l^2(m_1 k_2 + m_2 k_1) > 0, \\ b_3 &= a_0 a_4 = k_1 k_2 l^4 m_2 (m_1 + m_2 \sin^2\theta_0) > 0. \end{aligned}$$

On the other hand, the origin is unstable whenever $a_2 = b_1P - b_2 > 0$ (i.e., $\omega_i^2 < 0$). Thus, the critical value of the load for the flutter instability boundary is

$$P_{\text{cr}} = \frac{b_2 - 2\sqrt{b_3}}{b_1}.$$

In the following paragraphs, some illustrative numerical results are given. Table 4.2 lists the numerical value of the system parameters used in these simulations. Fig. 4.8 shows the evolution of the eigenvalues (ω^2) as the magnitude of the

Table 4.2 Sample parameter values

Parameter	Value	Parameter	Value
m_1	1 kg	k_1	1,000 N m/rad
m_2	1 kg	k_2	100 N m/rad
l	0.5 m	θ_0	$\pi/6$

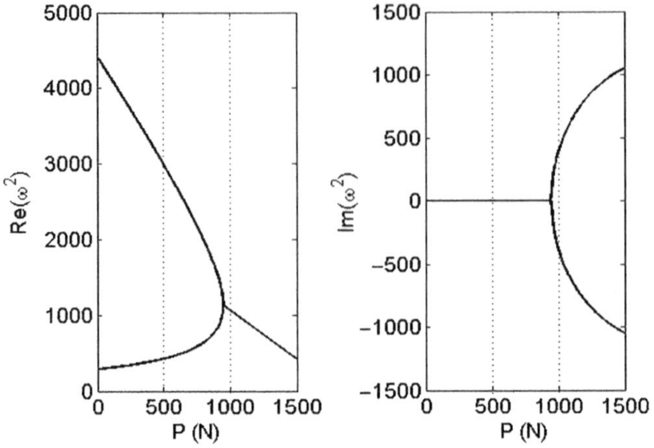

Fig. 4.8 Evolution of the real and imaginary parts of the eigenvalues of the undamped system

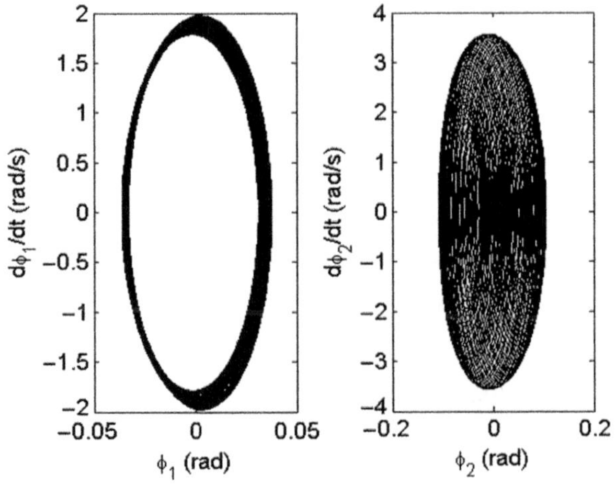

Fig. 4.9 Phase space projections of the nonlinear system for $P = 500$ N

follower force P is varied. At approximately $P = P_{\mathrm{cr}} \approx 900\,\mathrm{N}$, the two natural frequencies coalesce marking the flutter instability boundary. By increasing the force beyond this critical value, the eigenvalues become complex and the equilibrium point losses its stability. In Fig. 4.9, the four dimensional system trajectory is projected into $\varphi_1 - \dot{\varphi}_1$ and $\varphi_2 - \dot{\varphi}_2$ planes for $P = 500 < P_{\mathrm{cr}}$. The origin is stable according to Definition 2.1.

The flutter unstable behavior of this system is illustrated by phase projections in Fig. 4.10 for $P = 1{,}000 > P_{\mathrm{cr}}$. Figure 4.11 shows superimposed snapshots of the manipulator configurations over a period of time for the same unstable conditions.

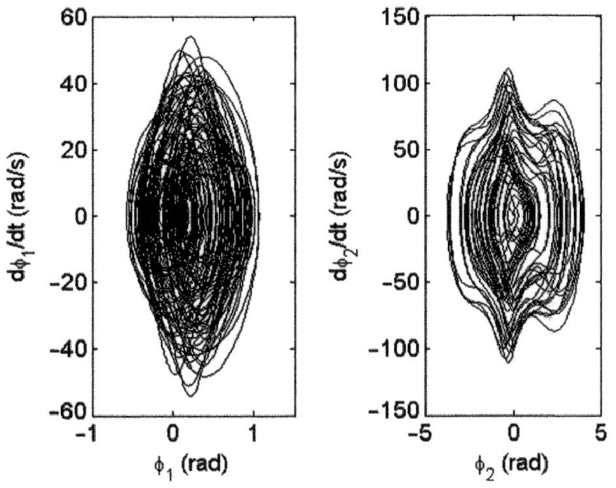

Fig. 4.10 Phase space projections of the nonlinear system for $P = 1{,}000$ N

Fig. 4.11 Flutter instability in the 2-DOF manipulator with follower force, $P = 1{,}000$ N

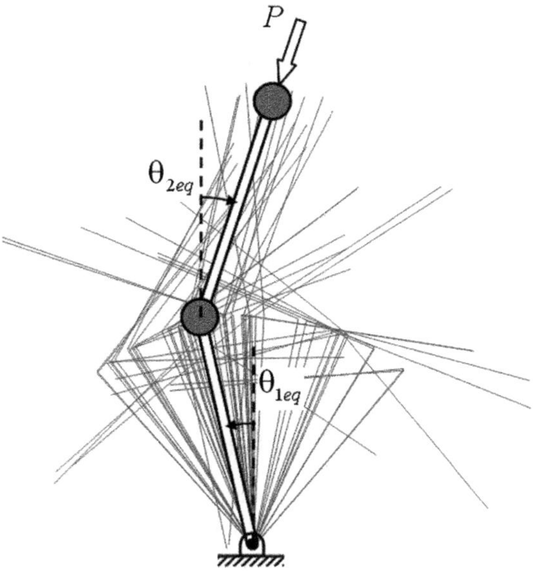

4.2.2 *Example No. 2: Mode Coupling*

Consider the 2-DOF system shown in Fig. 4.12 studied by Hoffman and Gaul [76, 77]. This model consists of a point mass sliding on a conveyor. The mass is suspended using vertical and horizontal linear springs and dampers. An additional spring placed at 45° angle is also considered which acts as the coupling between vertical

Fig. 4.12 A simple 2-DOF model capable of exhibiting mode coupling instability [76]

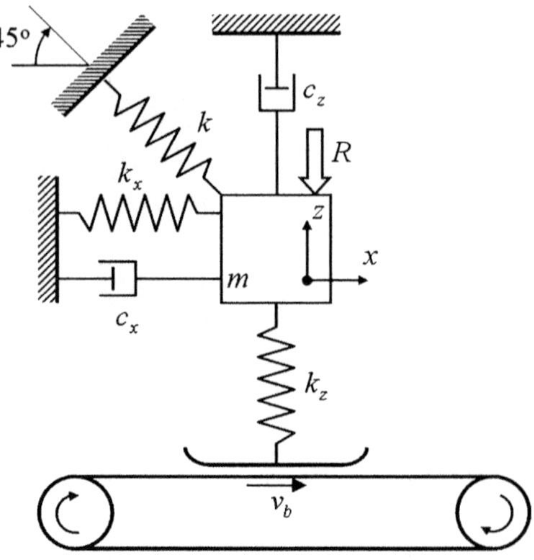

and horizontal motions. The friction force is modeled using Coulomb friction law; i.e., $F_\mathrm{t} = \mu F_\mathrm{n}$ where μ is the constant coefficient of friction. Also the conveyor belt is moving with constant velocity, $v_\mathrm{b} > 0$. The downward force R is assumed large enough to ensure that the contact between mass and conveyor belt is not lost.

The equation of motion for this system can be written in matrix form as

$$\mathbf{M}\ddot{\mathbf{q}} + \mathbf{C}\dot{\mathbf{q}} + \mathbf{K}\mathbf{q} = \widehat{\mathbf{f}}(\mathbf{q}, \dot{\mathbf{q}}),$$

where $\mathbf{q} = [x \quad z]^\mathrm{T}$ and $\mathbf{M} = \begin{bmatrix} m & 0 \\ 0 & m \end{bmatrix}, \mathbf{C} = \begin{bmatrix} c_x & 0 \\ 0 & c_z \end{bmatrix}$,

$$\mathbf{K} = \begin{bmatrix} k_x + \frac{1}{2}k & -\frac{1}{2}k \\ -\frac{1}{2}k & k_z + \frac{1}{2}k \end{bmatrix},$$

$$\mathbf{f}(\mathbf{q}, \dot{\mathbf{q}}) = \begin{bmatrix} -\mu k_z z\mathrm{sgn}(v_\mathrm{b} - \dot{x}) \\ -R \end{bmatrix}.$$

Shifting the equilibrium point (steady-sliding state) to the origin by setting $x_1 = x - x_\mathrm{eq}$ and $z_1 = z - z_\mathrm{eq}$, where

$$\begin{bmatrix} x_\mathrm{eq} \\ z_\mathrm{eq} \end{bmatrix} = \begin{bmatrix} k_x + \frac{1}{2}k & -\frac{1}{2}k + \mu k_z \\ -\frac{1}{2}k & k_z + \frac{1}{2}k \end{bmatrix}^{-1} \begin{bmatrix} 0 \\ -R \end{bmatrix}$$

gives

$$\mathbf{M}\ddot{\mathbf{y}} + \mathbf{C}\dot{\mathbf{y}} + \Lambda_\mathrm{p}\mathbf{y} = \mathbf{f}(\mathbf{y}, \dot{\mathbf{y}}),$$

where $\mathbf{y} = [y_1 \quad y_2]^{\mathrm{T}} = \mathbf{q} - \mathbf{q}_{\mathrm{eq}}$, $\mathbf{q}_{\mathrm{eq}} = [x_{\mathrm{eq}} \quad z_{\mathrm{eq}}]^{\mathrm{T}}$, and

$$\mathbf{\Lambda}_{\mathrm{p}} = \begin{bmatrix} k_x + \frac{1}{2}k & -\frac{1}{2}k + \mu k_z \\ -\frac{1}{2}k & k_z + \frac{1}{2}k \end{bmatrix}, \tag{4.38}$$

$$\mathbf{f}(\mathbf{y}, \dot{\mathbf{y}}) = \begin{bmatrix} \mu k_z(y_2 + z_0)\,[1 - \mathrm{sgn}(v_{\mathrm{b}} - \dot{y}_1)] \\ 0 \end{bmatrix}.$$

The symmetry-breaking role of friction is clearly shown by (4.38). Note that $\mathbf{f}(\mathbf{y}, \dot{\mathbf{y}})$ is nonzero only when $\dot{y}_1 \geq v_{\mathrm{b}}$. In a small neighborhood of the origin, (4.38) simplifies to a linear homogeneous differential equation

$$\mathbf{M}\ddot{\mathbf{y}} + \mathbf{C}\dot{\mathbf{y}} + \mathbf{\Lambda}_{\mathrm{p}}\mathbf{y} = \mathbf{0}, \quad |y_1| < v_{\mathrm{b}}. \tag{4.39}$$

Neglecting damping, from (3.29) the characteristic equations is found as

$$m^2\omega^4 - m(k_x + k_z + k)\omega^2 + k_x k_z + \tfrac{1}{2}k_x k + \tfrac{1}{2}(1 + \mu)kk_z = 0.$$

From (3.30), flutter instability threshold is calculated as

$$\mu_{\mathrm{cr}} = \frac{(k_x - k_z)^2 + k^2}{k_z k}.$$

If $\mu = \mu_{\mathrm{cr}}$, the two natural frequencies become identical, given by

$$\omega_1^2 = \omega_2^2 = \sqrt{\frac{k_x + k_z + k}{2m}}. \tag{4.40}$$

Increasing the coefficient of friction beyond its flutter critical value (i.e., $\mu > \mu_{\mathrm{cr}}$) results in a pair of complex conjugate squared natural frequencies, which indicates instability of the steady-sliding equilibrium point. Some illustrative numerical results are given next. Table 4.3 lists the numerical value of the system parameters used in these simulations.

We start by examining the undamped case (i.e., we set $c_x = c_z = 0$). Similar to Fig. 4.8, Fig. 4.13 shows the evolution of the eigenvalues of the undamped system.

Table 4.3 Sample parameter values

Parameter	Value	Parameter	Value
m	1 kg	k_x	100 N/m
k	100 N/m	k_z	200 N/m
v_{b}	10 m/s		

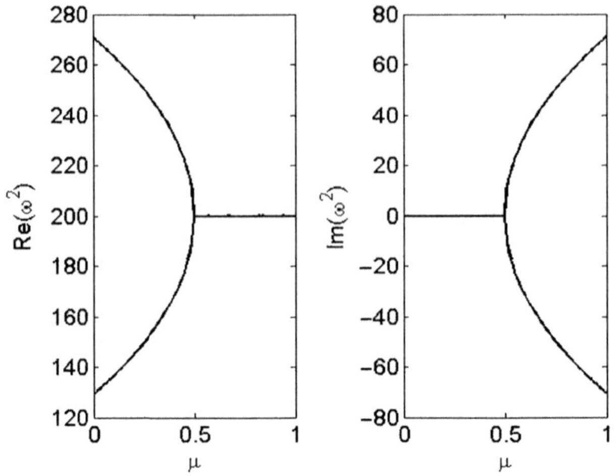

Fig. 4.13 Evolution of the real and imaginary parts of the eigenvalues of the undamped system

At the boundary of the flutter instability, $\mu = \mu_{\mathrm{cr}} = 0.5$, the two eigenvalues are identical and by further increasing the coefficient of friction, the eigenvalues become complex numbers.

Next we add damping by setting $c_x = 1.33$ N s/m and $c_z = 1$ N s/m. Generally, when damping is present, similar coalescence of the eigenvalues as in the undamped case is not observed.[4] This is certainly evident from the plot of variation of the real and imaginary parts of the eigenvalues in Fig. 4.14. In this example, the critical value of the coefficient of friction (i.e., flutter instability boundary) is $\mu_{\mathrm{cr}} \approx 0.52$.

The projections of the system trajectory onto $y_1 - \dot{y}_1$ and $y_2 - \dot{y}_2$ planes are shown in Fig. 4.15. The system trajectory is attracted to a limit cycle which touches the stick–slip boundary of $\dot{y}_1 = v_{\mathrm{b}}$ in the $y_1 - \dot{y}_1$ plane.

4.2.3 *Further References on Mode Coupling*

Recently, a great number of papers were published on the systems exhibiting mode coupling instability due to friction and the complex effect of damping on such systems. See papers by Hoffmann and his coworkers [79–81] and Jézéquel and his coworkers [82–89]. Other recent works on this subject include [90–94].

[4] Matching of the frequencies of the two coupled modes is exact for the special case of proportional damping (see, e.g., [78]).

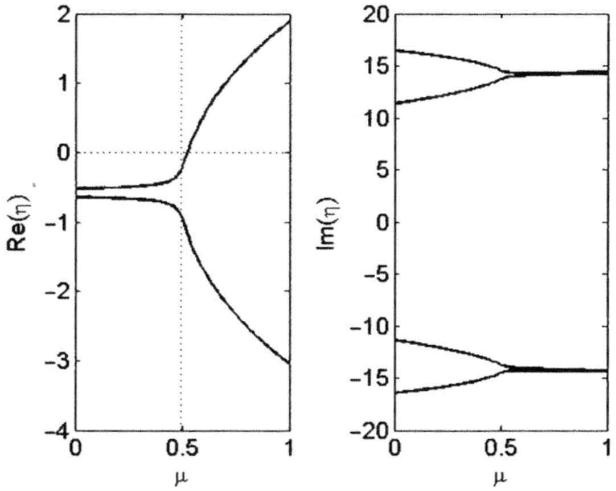

Fig. 4.14 Evolution of the real and imaginary parts of the eigenvalues of the damped system

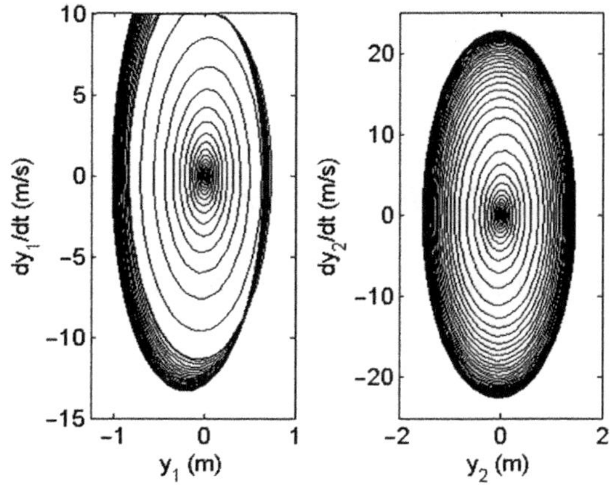

Fig. 4.15 Phase space projections of the nonlinear system for $\mu = 0.6$

4.3 Kinematic Constraint Instability

In Sect. 3.2, we have seen that a dynamical system with unilateral or bilateral frictional contact can possess a peculiar characteristic, namely the inertia matrix may be asymmetric and nonpositive definite. Painlevé was the first to point out the difficulties that may arise in such cases [53, 95]. As we will see in this section through examples, the presence of a kinematic constraint with friction could lead to situations where the equations of motion of the system do not have a bounded solution (inconsistency) or the solution is not unique (indeterminacy). These situations where the existence and uniqueness properties of the solution of the equations of motion are violated are known as the Painlevé's paradoxes. There is a vast literature on the general theory of the rigid body dynamics with frictional constraints (see, e.g., [96] and the references therein). Two of the focal points of the study of rigid bodies with friction are the determination of the conditions under which paradoxes occur; and determination of the true system response in the paradoxical regions. A thorough treatment of these topics and others in the field of rigid body dynamics is certainly beyond the scope of this book. However, we are interested in the paradoxes since the otherwise stable trivial equilibrium point of a system loses stability when paradoxes occur (see, e.g., [6, 7] and references therein). This phenomenon constitutes the last of the three friction-induced instability mechanisms we intend to study in the lead screw drives.

In Sects. 4.3.1 and 4.3.2, we study the classic Painlevé's example and derive the conditions for the occurrence of the paradoxes. In Sect. 4.3.3, the concept of "self-locking" is introduced which is closely related to the kinematic constraint instability mechanism. In the rigid body systems, this phenomenon is sometimes known as "jamming" or "wedging" [97]. As we will see later on, the self-locking is an important aspect of the study of the dynamics of the lead screws. In Sect. 4.3.4, a simple model of a vibratory system is analyzed where the kinematic constraint mechanism leads to instability. In the study of disc brake systems, similar instability mechanism is sometimes referred to as "sprag-slip" vibration [7]. Some further references are given in Sect. 3.3.5.

4.3.1 Painlevé's Paradox

Consider the system shown in Fig. 4.16. A bar of length l is in contact with a rough rigid surface at an angle θ. The equations of motion for this system are written as follows:

$$
\begin{aligned}
m\ddot{x} &= F_{\mathrm{f}}, \\
m\ddot{y} &= N - mg, \\
I\ddot{\theta} &= \frac{l}{2}(-F_{\mathrm{f}}\sin\theta - N\cos\theta),
\end{aligned}
\tag{4.41}
$$

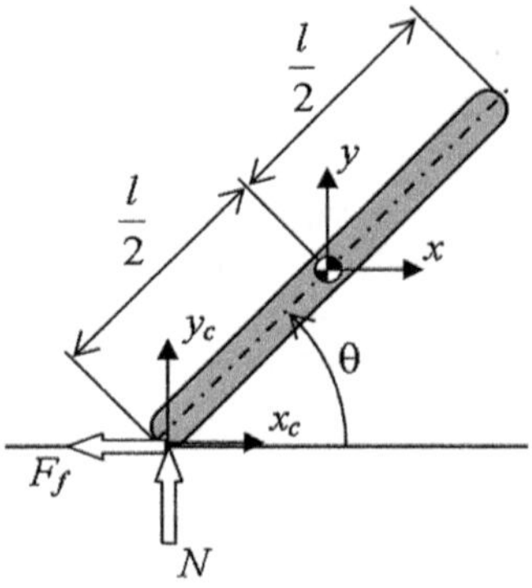

Fig. 4.16 The Painlevé's example

where m is the mass of the rod, I is its moment of inertia with respect to the center of mass, N is the normal contact force, and the friction force is given by

$$F_{\mathrm{f}} = \mu_{\mathrm{s}} N,$$

where the abbreviation

$$\mu_{\mathrm{s}} = \mu \mathrm{sgn}(\dot{x}_{\mathrm{c}})$$

is used. Also, μ is the constant coefficient of friction and $\dot{x}_{\mathrm{c}}$ is the tangential contact velocity. The position of the contacting tip of the rod is given by

$$y_{\mathrm{c}} = y - \frac{l}{2}\sin\theta. \tag{4.42}$$

When the rod is in contact with the surface we have $N \geq 0$ and $y_{\mathrm{c}} = 0$. On the other hand, when the rod brakes contact we have $N = 0$ and $y_{\mathrm{c}} \geq 0$. This situation can be represented as a *linear complementarity problem* [52, 96] which is written compactly as

$$0 \leq \ddot{y}_{\mathrm{c}} \perp N \geq 0. \tag{4.43}$$

Differentiating (4.42) twice with respect to time gives

$$\ddot{y}_{\mathrm{c}} = \ddot{y} - \frac{l}{2}\ddot{\theta}\cos\theta + \frac{l}{2}\dot{\theta}^2\sin\theta. \tag{4.44}$$

Substituting (4.41) into (4.44) yields

$$\ddot{y}_{\mathrm{c}} = AN + B,$$

where

$$\begin{aligned} A &= \frac{1}{m} + \frac{l^2}{4I}(\mu_{\mathrm{s}}\sin\theta + \cos\theta)\cos\theta, \\ B &= \frac{l}{2}\dot{\theta}^2\sin\theta - g. \end{aligned} \tag{4.45}$$

The solutions to the linear complementarity problem (4.43) can be represented graphically as shown in Fig. 4.17. As it can be seen from this figure, when $A < 0$ and $B > 0$, the solution is not unique and when $A < 0$ and $B < 0$ no solution exist.

Notice that the two necessary (but not sufficient) conditions for the paradoxes (i.e., $A < 0$) are

$$\mu > \cot\theta \text{ and } \mathrm{sgn}(\dot{x}_c) = -1.$$

4.3.2 Bilateral Contact

In the model shown in Fig. 4.18, the unilateral contact of model in Fig. 4.16 is replaced by a bilateral contact configuration. Here, the equation for the contact position is simply $y_c = 0$. From (4.42), the kinematic constraint equation is found as

$$y = \frac{l}{2}\sin\theta. \tag{4.46}$$

Also, the friction force is now given by

$$F_f = \mu|N|\text{sgn}(\dot{x}_c) = \mu_s N, \tag{4.47}$$

where

$$\mu_s = \mu\,\text{sgn}(\dot{x}_c)\text{sgn}(N). \tag{4.48}$$

Substituting (4.46) into (4.41) and using (4.47), the equation of motion of the system in y direction is found as

$$A\ddot{\theta} = \frac{l}{2I}B(\mu_s\sin\theta + \cos\theta),$$

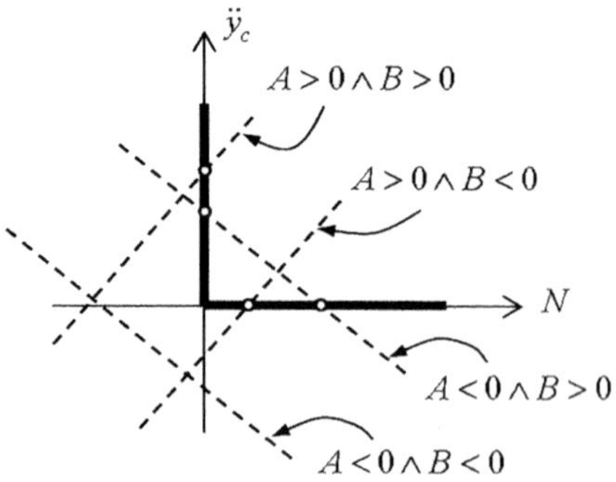

Fig. 4.17 Solutions of the complementarity problem

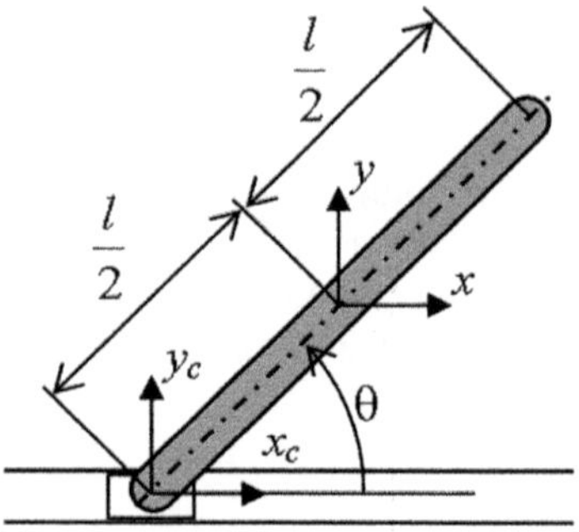

Fig. 4.18 Painlevé's example with bilateral contact

where A and B are given by (4.45). Note that μ_s is given by (4.48). Also, the contact force, N, is calculated from the following equation

$$N = \frac{B}{A}. \tag{4.49}$$

Similar to the case of the previous section, the nonexistence and nonuniqueness of the solution occur here if $\mathrm{sgn}(\dot{x}_c) = -1$ and $A < 0$:

If $B < 0$, two solutions are found for (4.49); setting $\mathrm{sgn}(N) = +1$ in the LHS of (4.49) results in

$$N_+ = \frac{(l/2)\dot{\theta}^2 \sin\theta - g}{(1/m) + (l^2/4I)(-\mu \sin\theta + \cos\theta)\cos\theta} > 0.$$

On the other hand, setting $\mathrm{sgn}(N) = -1$ gives

$$N_- = \frac{(l/2)\dot{\theta}^2 \sin\theta - g}{(1/m) + (l^2/4I)(\mu \sin\theta + \cos\theta)\cos\theta} < 0.$$

If $B > 0$, no valid solutions are found for (4.49); setting $\mathrm{sgn}(N) = +1$ in the LHS of (4.49) results in a *negative* contact force and setting $\mathrm{sgn}(N) = -1$ results in a *positive* contact force.

It is interesting to note that, for the parameter values that existence and uniqueness of solution are violated, the system's apparent inertia, *ImA*, is *negative*.

4.3.3 Self-Locking

Another consequence of friction in the dynamical systems is the possibility of self-locking (or self-breaking [51]). Consider the system shown in Fig. 4.19 [7]. In this model, a massless rigid rod pivoted at point O is contacting a rigid moving plane.

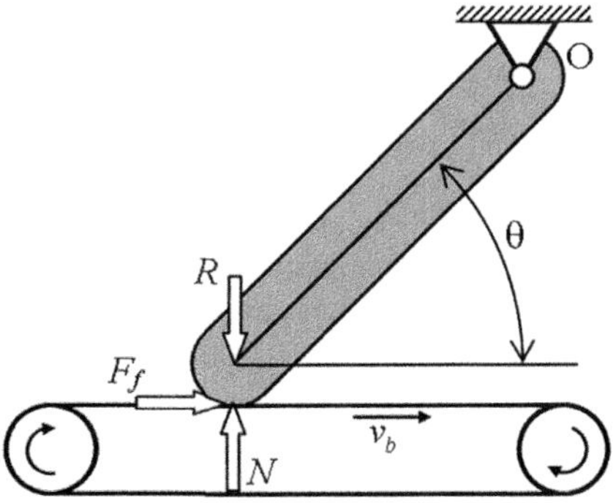

Fig. 4.19 Simple model to demonstrate kinematic constraint instability

A force R is pressing the free end of the rod against the moving plane. The normal and friction forces applied to the rod are given by N and $F_{\mathrm{f}} = \mu N$ where μ is the constant kinetic coefficient of friction. It can be shown that at equilibrium

$$N = \frac{R}{1 - \mu \tan\theta}. \tag{4.50}$$

From (4.50), it is evident that if $\theta \to \tan^{-1}(1/\mu_{\mathrm{k}})$, then $N \to \infty$ and further motion becomes impossible. In a more realistic setting where some flexibility is assumed, the motion continues by the deflection of the parts (see, e.g., Hoffmann and Gaul [98]). After sufficient deformation of the contacting bodies, slippage occurs which allows the bodies to assume their original configuration and the cycle continues. This situation is sometimes known as the sprag-slip limit cycle.

In the example treated in the next section, a similar model is considered with the exception of the addition of vertical compliance to the pivot location (and also to the contact).

4.3.4 An Example of Kinematic Constraint Instability

4.3.4.1 Mathematical Model

Consider the model shown in Fig. 4.20 which is similar to Fig. 4.19. Here, a rod of length l with mass m and moment of inertia I is pivoted at point O at one end and slides against a moving surface at another. Initially, the rod makes an angle θ_0 with

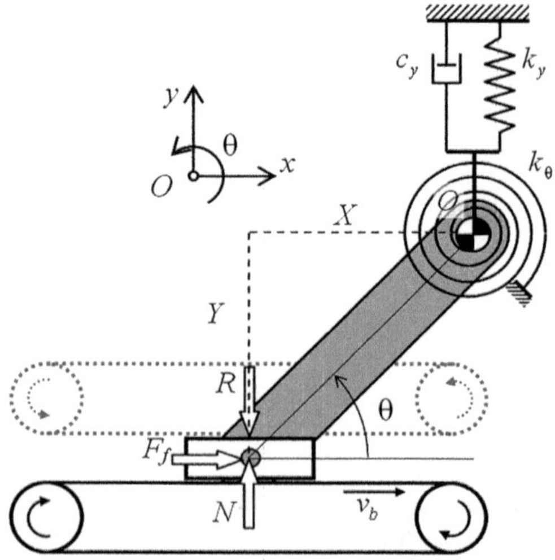

Fig. 4.20 Simple model to demonstrate sprag-slip vibration

the x-axis. Unlike the previous example, the joint at point O is given vertical compliance with linear spring (k_y) and linear damper (c_y). A torsional stiffness, k_θ, and torsional damping, c_θ, are also added to this joint.

Considering the degrees of freedom, y and θ (with respect to the point O as shown in Fig. 4.20), the equations of motion for this system can be written as (assuming bilateral contact between the end of the rod and the conveyor)

$$\begin{aligned} m\ddot{y} &= N - P_y - R, \\ I\ddot{\theta} &= F_f Y + (R - N)X - T, \end{aligned} \tag{4.51}$$

where N is the normal contact force, R is a small extra downward force. F_f is the friction force and it is calculated here as

$$F_f = \mu|N|\text{sgn}(v_b - v_t), \tag{4.52}$$

where μ is the constant coefficient of friction, $v_b > 0$ is the constant velocity of the conveyor's surface, and v_t is the horizontal velocity of the contacting tip of the rod which is given by

$$v_t = l\dot{\theta}\sin\theta. \tag{4.53}$$

Also, P_y is the vertical reaction force of the vertical linear spring and damper connected to the point O and is calculated as

$$P_y = ky + c\dot{y}, \tag{4.54}$$

T is the torque reaction of the rotational spring and damper at point O and it is calculated as

$$T = k_\theta(\theta - \theta_0) + c_\theta\dot{\theta}.$$

The bilateral contact between the rod and the moving surface, introduces the constraint:

$$y - l\sin\theta = 0. \tag{4.55}$$

Also, the moment arms X and Y are calculated as

$$\begin{aligned} Y &= l\sin\theta, \\ X &= l\cos\theta. \end{aligned} \tag{4.56}$$

Substituting (4.52), (4.54), and (4.56) into (4.51) gives

$$\begin{aligned} m\ddot{y} + k_y y + c_y \dot{y} &= N - R, \\ I\ddot{\theta} + k_\theta \theta + c_\theta \dot{\theta} &= k_\theta \theta_0 + \mu_s N l \sin\theta + (R - N) l \cos\theta, \end{aligned} \tag{4.57}$$

where the abbreviations

$$\mu_s = \mu \mathrm{sgn}(v_t)\mathrm{sgn}(N) \tag{4.58}$$

and

$$v_t = v_b - l\dot{\theta}\sin\theta$$

were used. Eliminating N between the two equations of (4.57) and using the *constraint equation* (4.55) give the equation of motion of the rod rotation, θ. After some algebra, one finds

$$M(\theta)\ddot{\theta} + C(\theta)\dot{\theta} + G(\theta)\dot{\theta}^2 + F(\theta) = 0, \tag{4.59}$$

where

$$\begin{aligned} M(\theta) &= I + ml^2\cos^2\theta\xi(\theta), \\ C(\theta) &= c_y l^2 \cos^2\theta\xi(\theta) + c_\theta, \\ G(\theta) &= -ml^2 \sin\theta\cos\theta\xi(\theta), \\ F(\theta) &= k_y l^2 \sin\theta\cos\theta\xi(\theta) + k_\theta(\theta - \theta_0) - lR\mu_s \sin\theta. \end{aligned} \tag{4.60}$$

Also

$$\xi(\theta) = 1 - \mu_s \tan\theta.$$

Note that, M represents the system inertia, C is the nonlinear damping coefficient, and F_k represents the nonlinear elastic forces. G and F_R account for the effects of centrifugal and external forces, respectively.

The normal force, N, is found from (4.57) as

$$N = \frac{\Lambda\left(\theta, \dot{\theta}\right)}{M(\theta)}, \tag{4.61}$$

where

$$\begin{aligned} \Lambda\left(\theta, \dot{\theta}\right) &= Ik_y l \sin\theta - k_\theta ml \cos\theta(\theta - \theta_0) + \left(ml^2\cos^2\theta + I\right)R \\ &\quad - mIl\dot{\theta}^2 \sin\theta + \left(Ic_y - c_\theta m\right)l\dot{\theta}\cos\theta \end{aligned} \tag{4.62}$$

and $M(\theta)$ is given by (4.60). Similar to Sect. 4.3.2, the Painlevé's paradox is encountered if $M(\theta) < 0$. The necessary conditions for $M(\theta) < 0$ is

$$\mu > \cot\theta \text{ and } Nv_t > 0,$$

which are the same conditions as the self-locking in the example of the previous section.

The equation of motion given by (4.59) can have multiple equilibriums which are the solutions of $F(\theta_{eq}) = 0$. Here, we restrict ourselves to the case where there is only one equilibrium point. We assume that k_θ is large enough such that for $Nv_t > 0$,

$$K(\theta) = \frac{\partial F(\theta)}{\partial \theta} = k_\theta + k_y l^2 \cos 2\theta (1 - \mu \tan 2\theta) - lR\mu \cos\theta > 0, \quad 0 < \theta < \frac{\pi}{2}.$$

The equilibrium point is approximately found as (using Newton–Raphson method)

$$\theta_{eq} \approx \theta_0 - \frac{F(\theta_0)}{K(\theta_0)}.$$

The normal force at equilibrium is found from (4.61) to be $N_{eq} = k_y l \sin\theta_{eq} + R$ which is positive for $R > 0$ and $0 < \theta_{eq} < \pi$. The linearized equation of motion with respect to small motions around the equilibrium point can be written as

$$M(\theta_{eq})\ddot{\theta} + C(\theta_{eq})\dot{\theta} + K(\theta_{eq})\theta = 0. \tag{4.63}$$

4.3.4.2 Negative Damping Instability

Since $K(\theta_{eq}) > 0$, instabilities occur if either $M(\theta_{eq}) < 0$ (Painlevé's Paradox) or $C(\theta_{eq}) < 0$. (Negative damping). Starting with the negative damping instability, we can see that if the following conditions are satisfied, the origin of the system (4.63) becomes unstable.

$$\mu > \frac{c_\theta}{c_y l^2 \sin\theta_{eq} \cos\theta_{eq}} + \cot\theta_{eq}. \tag{4.64}$$

Note that if (4.64) is satisfied, the self-locking condition (i.e., $\mu > \cot\theta$) also holds. Next, some numerical simulation results are presented to illustrate the negative damping consequence of the kinematic constraint. The numerical value of the parameters are taken from Table 4.4.

Table 4.4 Sample parameter values

Parameter	Value	Parameter	Value
m	1 kg	μ	1.2
I	4 kg m^2	c_y	1 N/(m/s)
k_y	10 N/m	v_b	5 m/s
K_θ	1,000 N m/rad	θ_0	$\pi/4$ rad
R	1 N	l	5 m

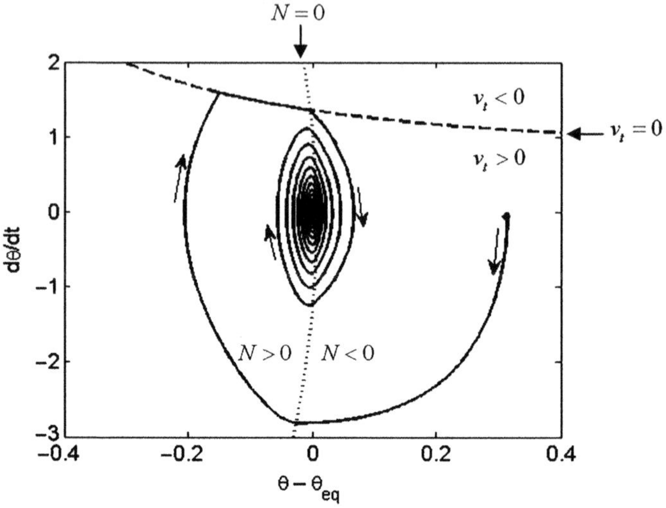

Fig. 4.21 Stable steady-sliding equilibrium point, C(θ_{eq}) $\approx$ 1.5

In Figs. 4.21 and 4.22, system trajectories are shown for $c_\theta = 5$ and $c_\theta = 0.5$ N m s/rad, respectively.

In these figures, the curve defined by $v_t(\theta, \dot{\theta}) = 0$, which indicates the stick boundary, and the curve defined by $N(\theta, \dot{\theta}) = 0$, which is the boundary where the contact force changes sign, are also shown.

For the selected values of the rotational damping, the effective linear damping coefficients are found as

$$c_\theta = 5 \rightarrow C(\theta_{eq}) \approx 1.5 > 0,$$
$$c_\theta = 0.5 \rightarrow C(\theta_{eq}) \approx -3 < 0.$$

As can be seen from Fig. 4.22, for $C(\theta_{eq}) < 0$ the steady-sliding equilibrium point is unstable and the trajectory is attracted to a limit cycle. Note that, in these two cases, the system inertia – found from (4.60) – is $M(\theta_{eq}) \approx 0.49 > 0$ (i.e., no paradoxes).

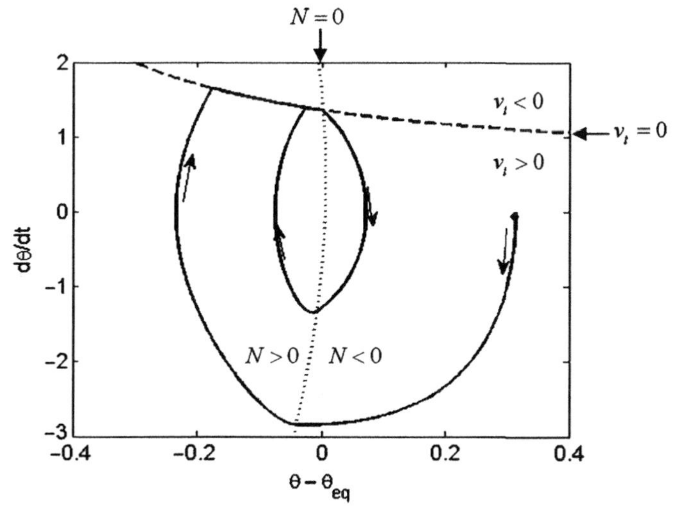

Fig. 4.22 Unstable steady-sliding equilibrium point, $C(\theta_{eq}) \approx -3 < 0$

4.3.4.3 Painlevé's Paradoxes

Consider the following two cases for the system's effective inertia:

$$M(\theta) = \begin{cases} M_+(\theta) = I + ml^2\cos^2\theta(1 - \mu\tan\theta), & N \cdot v_t > 0 \\ M_-(\theta) = I + ml^2\cos^2\theta(1 + \mu\tan\theta), & N \cdot v_t < 0 \end{cases}, \quad 0 < \theta < \frac{\pi}{2}. \tag{4.65}$$

$M_-(\theta_{eq})$ is always positive, however, $M_+(\theta_{eq})$ becomes negative for friction values satisfying

$$\mu \geq \frac{I\cot\theta_{eq}}{ml^2\sin\theta_{eq}\cos\theta_{eq}} + \cot\theta_{eq}.$$

For the parameter values that satisfy the above inequality and $Nv_t < 0$, Painlevé's paradoxes occur and the equation of motion given by (4.59) or the linearized equation given by (4.63) are no longer valid due to violation of existence and uniqueness of the solution. However, as we will see in the numerical examples below, the steady-sliding equilibrium point is indeed *unstable* for such values of system parameters.

Figure 4.23 shows a portion of the phase plane of the system for the parameters given in Table 4.5.

For these parameter values, we have $M(\theta_{eq}) \approx -6.4 < 0$ which indicates the occurrence of the Painlevé's paradoxes. Note that, here we have $C(\theta_{eq}) \approx 4.8 > 0$. Three curves divide the phase plane into five regions: vertical line $M_+(\theta) = 0$; stick boundary $v_t(\theta, \dot{\theta}) = 0$; and the curve defined by $\Lambda(\theta, \dot{\theta}) = 0$ where $\Lambda(\theta, \dot{\theta})$ is given by (4.62).

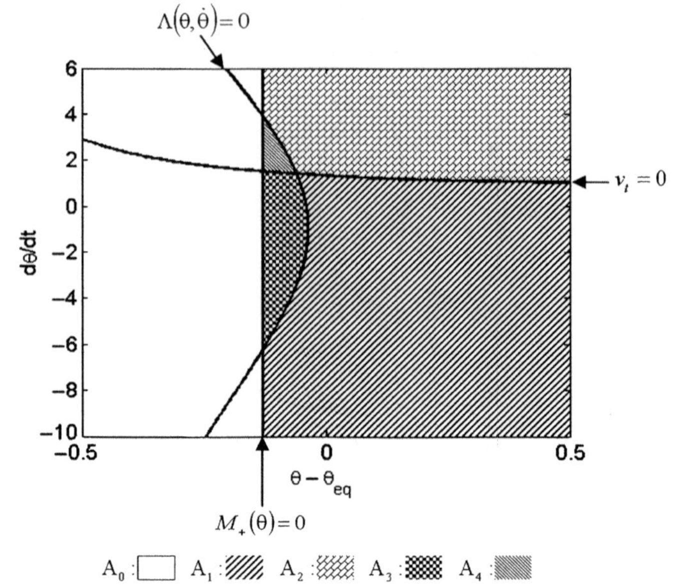

Fig. 4.23 Regions of paradoxes in the system's phase plane

Table 4.5 Sample parameter values

Parameter	Value	Parameter	Value
m	2 kg	c_y	1 N/(m/s)
I	4 kg m^2	c_θ	10 N/(m/s)
k_y	10 N/m	v_b	10 m/s
k_θ	1,000 N m/rad	θ_0	$\pi/4$ rad
R	1 N	l	5 m
μ	1.3		

Table 4.6 Number of solutions in the five regions of Fig. 4.23

Region	$M_+(\theta)$	v_t	$\Lambda(\theta, \dot{\theta})$	Number of solutions
A_0	+	+/−	+/−	1
A_1	−	+	−	2
A_2	−	−	−	None
A_3	−	+	+	None
A_4	−	−	+	2

As listed in Table 4.6, four of the five regions identified in Fig. 4.23 correspond to initial states where either no solution exists or it is not unique. As mentioned above, paradoxes does not occur when $M_+(\theta) > 0$ (region A_0). In regions where

$M_+(\theta) < 0$, there are two possibilities: $\Lambda(\theta,\dot{\theta})\cdot v_t(\theta,\dot{\theta}) > 0$ which leads to inconsistency; and $\Lambda(\theta,\dot{\theta}) \cdot v_t(\theta,\dot{\theta}) < 0$ which leads to indeterminacy.

If $M_+ < 0$, $\Lambda > 0$, $v_t > 0$, and assuming $N > 0$ from (4.61), we find

$$N = \frac{\Lambda}{M|_{Nv_t > 0}} = \frac{\Lambda > 0}{M_+ < 0} < 0$$

and if $N < 0$,

$$N = \frac{\Lambda}{M|_{Nv_t < 0}} = \frac{\Lambda > 0}{M_- > 0} > 0.$$

Thus, the normal force equation (4.61) has no solution in region A_3. If $M_+ < 0$, $\Lambda < 0$, $v_t < 0$, and assuming $N > 0$ from (4.61), we find

$$N = \frac{\Lambda}{M|_{Nv_t < 0}} = \frac{\Lambda < 0}{M_- > 0} < 0$$

and if $N < 0$,

$$N = \frac{\Lambda}{M|_{N\cdot v_t > 0}} = \frac{\Lambda < 0}{M_+ < 0} > 0.$$

Thus, the normal force equation (4.61) has no solution region A_2. If $M_+ < 0$, $\Lambda < 0$, $v_t > 0$, and assuming $N > 0$ from (4.61), we find

$$N = \frac{\Lambda}{M|_{N\cdot v_t > 0}} = \frac{\Lambda < 0}{M_+ < 0} > 0$$

and if $N < 0$,

$$N = \frac{\Lambda}{M|_{N.v_t < 0}} = \frac{\Lambda < 0}{M_- > 0} < 0.$$

Thus, the normal force equation (4.61) has two solutions in region A_1. Finally, if $M_+ < 0$, $\Lambda > 0$, $v_t < 0$, and assuming $N > 0$ from (4.61), we find

$$N = \frac{\Lambda}{M|_{N\cdot v_t < 0}} = \frac{\Lambda > 0}{M_- > 0} > 0$$

and if $N < 0$,

$$N = \frac{\Lambda}{M|_{N\cdot v_t > 0}} = \frac{\Lambda > 0}{M_+ < 0} < 0.$$

Thus, the normal force equation (4.61) has two solutions in region A_4. Similar to (4.65), we define

$$C(\theta) = \begin{cases} C_+(\theta) = c_y l^2 \cos^2\theta(1-\mu\tan\theta) + c_\theta, & Nv_t > 0 \\ C_-(\theta) = c_y l^2 \cos^2\theta(1+\mu\tan\theta) + c_\theta, & Nv_t < 0 \end{cases},$$

$$G(\theta) = \begin{cases} G_+(\theta) = -ml^2 \sin\theta\cos\theta(1-\mu\tan\theta), & Nv_t > 0 \\ G_-(\theta) = -ml^2 \sin\theta\cos\theta(1+\mu\tan\theta), & Nv_t < 0 \end{cases},$$

$$F(\theta) = \begin{cases} F_+(\theta) = k_y l^2 \sin\theta\cos\theta(1-\mu\tan\theta) + k_\theta(\theta-\theta_0) - lR\mu\sin\theta, & Nv_t > 0 \\ F_-(\theta) = k_y l^2 \sin\theta\cos\theta(1+\mu\tan\theta) + k_\theta(\theta-\theta_0) + lR\mu\sin\theta, & Nv_t < 0 \end{cases}.$$

For the initial conditions in either regions A_1 or A_4, the two possible solutions are found from:

$$\begin{aligned} M_+(\theta)\ddot{\theta} + C_+(\theta)\dot{\theta} + G_+(\theta)\dot{\theta}^2 + F_+(\theta) = 0, \\ M_-(\theta)\ddot{\theta} + C_-(\theta)\dot{\theta} + G_-(\theta)\dot{\theta}^2 + F_-(\theta) = 0. \end{aligned} \tag{4.66}$$

Note that the origin is an unstable equilibrium point of the first system in (4.66) and a stable equilibrium point of the second one. In the next section, we use a compliant contact model to investigate the system's motion in the regions of paradoxes.

4.3.4.4 Motion in the Region of Paradoxes: Compliant Contact Model

To analyze the behavior of a system in the paradoxical regions, one way is to give the rigid contact some degree of compliance [99, 100]. Consider the system in Fig. 4.24. The equations of motion of this 2-DOF system are given by (4.57). With the additional equation for the normal contact force given by

$$N = -k_c(y - l\sin\theta) - c_c\left(\dot{y} - l\dot{\theta}\cos\theta\right),$$

where k_c and c_c are the linear stiffness and damping coefficients of the contact, respectively. Note that, similar to the model in the previous section, we assume bilateral contact between the slider and the conveyor.

For the parameter values given in Table 4.5 and $k_c = c_c = 10^6$, Fig. 4.25 shows a number of system trajectories. Comparing this figure with Fig. 4.23, one can see that for the trajectories that enter either region A_2 or region A_3 (inconsistency), dynamic seizure occurs and the motion restarts from the stick–slip boundary (see vertical lines in these regions). In the limit, as the contact stiffness is increased

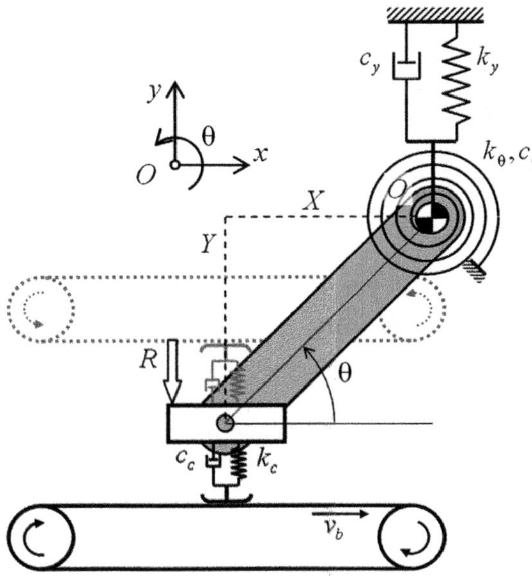

Fig. 4.24 Adding compliance to the contact point

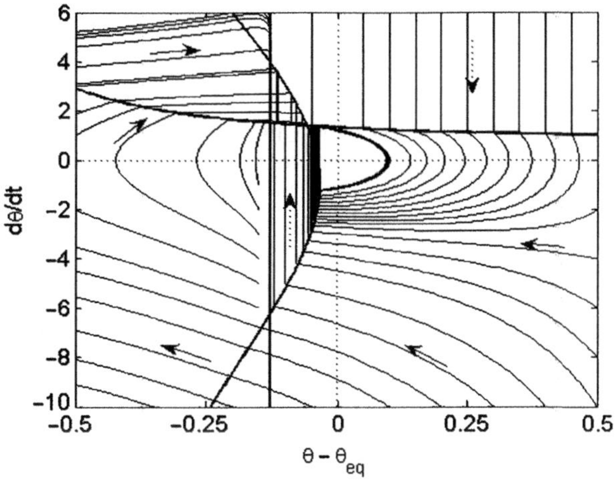

Fig. 4.25 Instability caused by Painlevé's paradox

to infinity, this effect is known as tangential impact or impact without collision (IW/OC) [51].

For the trajectories in either region A_1 or region A_4 (indeterminacy), the above numerical results show that the solution corresponds to $Nv_t < 0$. Note that, in the limit (as contact compliance tends to infinity), this solution converges to the solution of the second differential equation in (4.66).

For the 2-DOF system of Fig. 4.24, the initial conditions can be determined such that the initial contact force is any given value. It can be shown that there are initial values for the normal force where the solutions in either region A_1 or region A_4 correspond to $Nv_t > 0$ (not shown in the figure). In the limit, these solutions also converge to an impulsive solution for the contact force under which seizure occurs [51].

4.3.5 Further References on the Kinematic Constraint Instability Mechanism

Our study of the lead screw drives entails systems with a single bilateral contact with friction (between lead screw threads and nut threads). As demonstrated by the example in Sect. 4.3.4, in order to study the behavior of a system in the paradoxical regions of parameters, a compliant approximation to rigid contact may be used. In Chap. 8, the limit process approach presented in [51] is utilized to determine the true motion of a 1-DOF lead screw drive model under similar paradoxical conditions. In the limit process approach, the behavior of the rigid body system is taken as that of a similar system with compliant contacts when the contact stiffness tends to infinity. Related to this topic, a discussion of the method of penalizing function can be found in Brogliato ([96], Chap. 2). Other examples include [101–103].

The area of rigid body dynamical systems with contact and friction belongs to the study of nonsmooth systems. See, for example, [52, 104–108] for the theory of nonsmooth mechanics. Mathematical concepts, such as Filippov systems, measure differential inclusions, and linear complimentarity problems (LCP) are used to describe and analyze these systems. The book by Brogliato [96] is an excellent reference on these subjects and discusses a great number of relevant works.

Recent works on the Painlevé's classical example – introduced in Sect. 4.3.1 – include [109–111]. Friction impact oscillator which is similar to the system studied in Sect. 4.3.4 but with unilateral contact (creating the possibility of detachment and flight phases) is the subject of many publications; see, for example, [53, 112, 113].

Chapter 5
Mathematical Modeling of Lead Screw Drives

In this chapter, a collection of mathematical models are developed which are used to study the dynamic behavior of lead screw systems in the subsequent chapters. Depending on the system elements considered and the type of analysis undertaken, different models are developed with varying number of degrees of freedom.

Figure 5.1 shows a typical lead screw drive system. A motor – possibly through a gearbox – rotates the lead screw via a coupling. The rotational motion is converted to translation at the lead screw–nut interface and transferred to the moving mass. The weight of the moving mass is supported by bearings. The lead screw is held in place by support bearings at its either ends.

The velocity-dependent friction model used in this work is discussed in Sect. 5.1. The dynamics of a pair of meshing lead screw and nut threads is studied in Sect. 5.2. Based on the relationships derived in this section, the basic 1-DOF lead screw drive model is developed in Sect. 5.3. This model is used in Chaps. 6 and 8 to study the negative damping and kinematic constraint instability mechanisms, respectively. A model of the lead screw with antibacklash nut is presented in Sect. 5.4, and the role of preloaded nut on the increased friction is highlighted. Additional DOFs are introduced to the basic lead screw model in Sects. 5.5 to 5.8 in order to account for the flexibility of the threads, the axial flexibility of the lead screw supports, and the rotational flexibility of the nut. These models are used in Chaps. 7 and 8 to investigate the mode coupling and the kinematic constraint instability mechanisms, respectively. Finally, in Sect. 5.9, some remarks are made regarding the models developed in this chapter.

The rubbing action of the contacting lead screw threads against the nut threads is assumed to be the main source of friction in the systems considered in this monograph. We start this chapter by presenting a velocity-dependent coefficient of friction model for the lead screw and nut interface.

O. Vahid-Araghi and F. Golnaraghi, *Friction-Induced Vibration in Lead Screw Drives*,
DOI 10.1007/978-1-4419-1752-2_5,

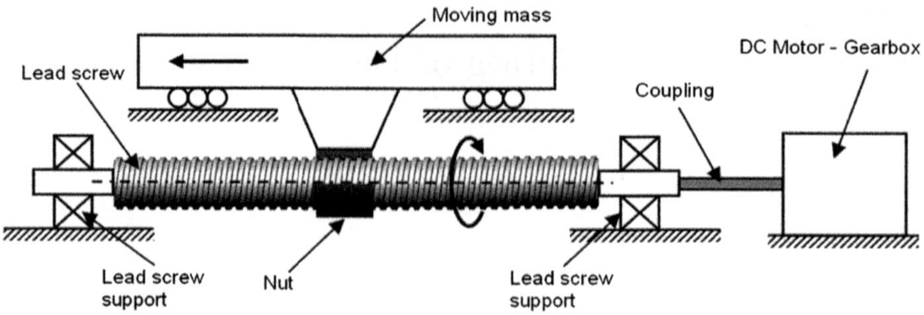

Fig. 5.1 Lead screw drive system

Fig. 5.2 Velocity-dependent coefficient of friction

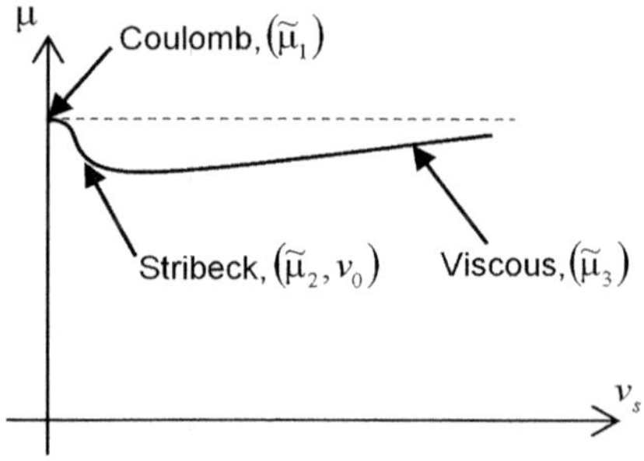

5.1 Velocity-Dependent Coefficient of Friction

As mentioned in Sect. 1.1, numerous models for the velocity-dependent coefficient of friction can be found in the literature [5, 11, 12]. These models generally include the following three parts:

1. Coulomb or constant friction
2. Stribeck friction
3. Viscous or linear friction

Here, the following model for the friction coefficient is considered[1]

$$\mu = \tilde{\mu}_1 + \tilde{\mu}_2(\mathrm{e}^{-(|v_\mathrm{s}|/v_0)} - 1) + \tilde{\mu}_3|v_\mathrm{s}|, \tag{5.1}$$

where $\tilde{\mu}_1$, $\tilde{\mu}_2$, and $\tilde{\mu}_3$ represent Coulomb, Stribeck, and viscous friction coefficients, respectively. v_s is the relative sliding velocity between contacting nut threads and lead screw threads. Also, v_0 controls the velocity range of the Stribeck effect. See Fig. 5.2 for a schematic plot of the velocity-dependent coefficient of friction given by (5.1).

[1]This model is sometimes known as the Tustin model [16].

The reasons for choosing this friction model are twofold. This model structure allows for the three above-mentioned components of friction to be easily separated for the purpose of focused analysis. In addition, it was found that this particular formula lends itself very well to the experimental observations reported in Chap. 9.

Based on Fig. 2.8, the sliding velocity can be written as

$$v_s = \frac{r_m}{\cos\lambda}\dot{\theta}. \tag{5.2}$$

Substituting (5.2) into (5.1) and rearranging,

$$\mu = \mu_1 + \mu_2 e^{-r_0|\dot{\theta}|} + \mu_3|\dot{\theta}|, \tag{5.3}$$

where $r_0 = r_m/\cos\lambda v_0$, $\mu_1 = \tilde{\mu}_1 - \tilde{\mu}_2$, $\mu_2 = \tilde{\mu}_2$, and $\mu_3 = \tilde{\mu}_3(r_m/\cos\lambda)$. In the sequel, (5.3) is used as the basic model for the velocity-dependent coefficient of friction.

5.2 Dynamics of Lead Screw and Nut

Figure 5.3 shows a pair of meshing *unwrapped* lead screw and nut threads. x designates nut travel and y given by

$$y = r_m\theta \tag{5.4}$$

is the equivalent translation of the lead screw (see Fig. 2.7). Also m_1 is the mass of the translating part and m_2 defined by

$$m_2 = \frac{I}{{r_m}^2} \tag{5.5}$$

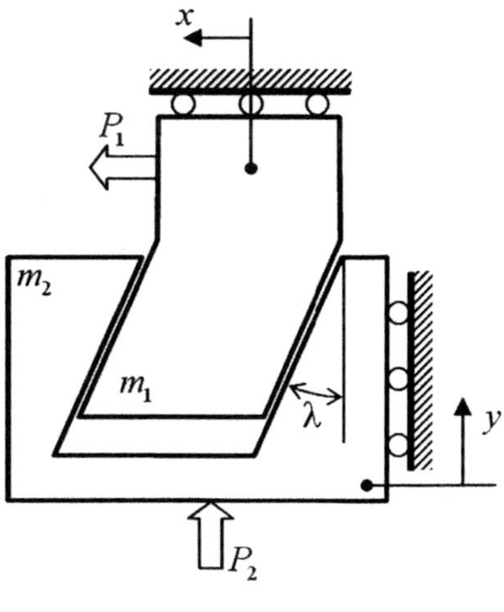

Fig. 5.3 Unwrapped lead screw and nut threads

is the equivalent lead screw mass where I is the lead screw moment of inertia. P_1 is the axial force applied to the nut and P_2 defined by

$$P_2 = \frac{T}{r_m} \tag{5.6}$$

is the equivalent force applied to the lead screw where T is the applied torque.

The equations of motion of this system (based on the notations of Fig. 2.7) can be written as

$$\begin{aligned} m_1\ddot{x} &= -N\cos\lambda - F_f\sin\lambda + P_1, \\ m_2\ddot{y} &= N\sin\lambda - F_f\cos\lambda + P_2. \end{aligned} \tag{5.7}$$

Based on (2.4), the friction force is written as

$$F_f = \mu_s N, \tag{5.8}$$

where

$$\mu_s = \mu\,\mathrm{sgn}(\dot{y}N). \tag{5.9}$$

Using (5.8) and (2.3) and eliminating N between the two equations in (5.7) yield

$$(m_2 - \xi\tan\lambda m_1)\ddot{y} = P_2 - \xi P_1, \tag{5.10}$$

where

$$\xi = \frac{\mu_s - \tan\lambda}{1 + \mu_s\tan\lambda}. \tag{5.11}$$

In addition, the contact normal force between the threads is found as

$$N = \frac{P_1 m_2 - P_2 m_1\tan\lambda}{(\cos\lambda + \mu_s\sin\lambda)(m_2 - \xi\tan\lambda m_1)}. \tag{5.12}$$

Substituting (5.4), (5.5), (5.6), into (5.7) yields

$$(I - \xi\tan\lambda m{r_m}^2)\ddot{\theta} = T - r_m\xi P, \tag{5.13}$$

where the index 1 is dropped to simplify the notation. Similarly, the contact force given by (5.12) becomes

$$N = \frac{PI - Tr_m m\tan\lambda}{(\cos\lambda + \mu_s\sin\lambda)(I - \xi\tan\lambda m{r_m}^2)}. \tag{5.14}$$

Note that due to the appearance of $\mathrm{sgn}(N)$, through μ_s, in the denominator of (5.14), this equation is solved iteratively for N.

In the remainder of this chapter, a number of lead screw models are presented which are based on the above derivations. For convenience – and with some abuse of symbols – the unwrapped threads pair depicted in Fig. 5.3 is used to represent a pair of meshing lead screw and nut throughout the rest of this monograph.

5.3 Basic 1-DOF Model

Figure 5.4 shows a 1-DOF lead screw drive model based on the model in Fig. 5.3.[2] θ_i is the input rotational displacement applied to the lead screw through a flexible coupling (torsional spring k). R is the axial force applied to the nut. Also, c is the rotational linear damping coefficient of the lead screw supports and c_x is the linear damping coefficient of the bearing supporting the translating part. T_0 and F_0 are the frictional torque and force of the translating part and the lead screw supports, respectively.

Based on the formulation of the previous section, the torque T and the force P in (5.13) are given as

$$T = k(\theta_i - \theta) - c\dot{\theta} - T_0 \operatorname{sgn}(\dot{\theta}),$$
$$P = R - c_x\dot{x} - F_0 \operatorname{sgn}(\dot{x}). \tag{5.15}$$

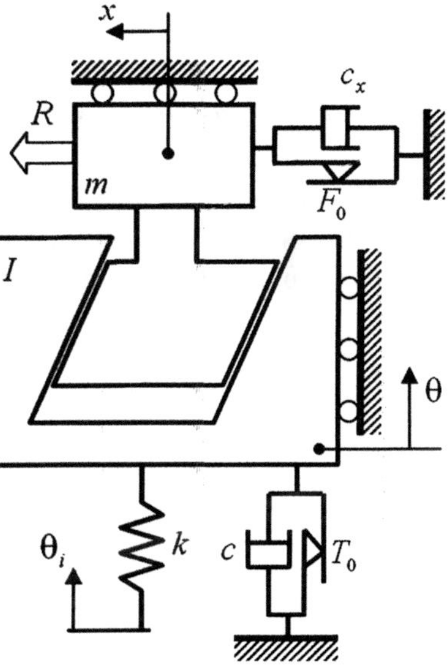

Fig. 5.4 1-DOF model of a lead screw system

[2]By suitably selecting conventions for the axes and forces directions, the equations of motion derived here apply to both left-handed and right-handed screws. As a result, from this point on, the handedness of the lead screw is assumed to be known but is not included in the discussions.

The equations obtained from the application of the Newton's second law to the lead screw and nut are repeated here for future reference. We have

$$I\ddot{\theta} = T + r_{\mathrm{m}}(N\sin\lambda - F_{\mathrm{f}}\cos\lambda), \tag{5.16}$$

$$m\ddot{x} = -N\cos\lambda - F_{\mathrm{f}}\sin\lambda + P. \tag{5.17}$$

Eliminating N between (5.16) and (5.17) and substituting (5.15) yield

$$(I - {r_{\mathrm{m}}}^2\tan\lambda\xi m)\ddot{\theta} + k\theta + (c - {r_{\mathrm{m}}}^2\tan\lambda\xi c_x)\dot{\theta}$$
$$= k\theta_{\mathrm{i}} - r_{\mathrm{m}}\xi(R - F_0\mathrm{sgn}(\dot{\theta})) - T_0\,\mathrm{sgn}(\dot{\theta}), \tag{5.18}$$

where (2.3), (5.8), and (5.15) are used. Substituting (5.15) into (5.14) yields the normal contact force as

$$N = \frac{(R - F_0\ \mathrm{sgn}(\dot{\theta}))I + mr_{\mathrm{m}}\tan\lambda\left[k(\theta - \theta_{\mathrm{i}}) + (c - {r_{\mathrm{m}}}^2\tan\lambda\xi c_x)\dot{\theta} + T_0\mathrm{sgn}(\dot{\theta})\right]}{(\cos\lambda + \mu_{\mathrm{s}}\sin\lambda)(I - {r_{\mathrm{m}}}^2\tan\lambda\xi m)}. \tag{5.19}$$

The equation of motion derived in this section can also describe other variations of the basic model which are discussed next. These models reflect other possible configurations that may be found in practice.

5.3.1 *Inverted Basic Model*

In some applications, the nut is rotated which causes the lead screw to translate. Figure 5.5 shows this configuration for a simple 1-DOF model. It can be shown that for this configuration, the equation of motion is identical to (5.18).

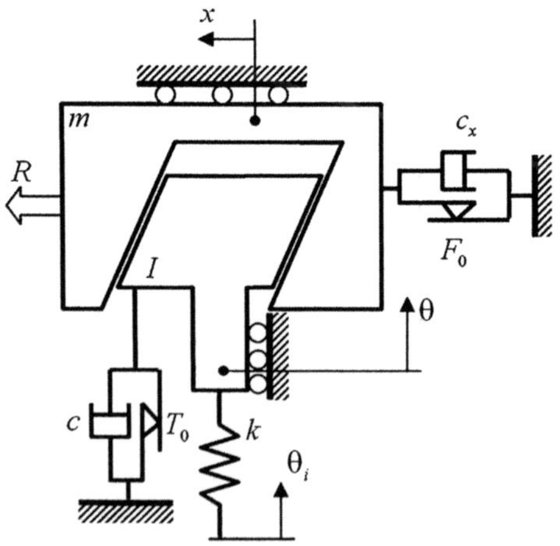

Fig. 5.5 Inverted basic 1-DOF model

5.3.2 *Basic Model with Fixed Nut*

In another possible configuration, the nut may be fixed and the lead screw rotation is converted to its translation together with other connected parts (i.e., motor, frame, payload, etc.). This configuration is shown in Fig. 5.6.

The equation of motion of this system is also given by (5.18).

5.3.3 *Basic Model with Fixed Lead Screw*

The last variation of the basic lead screw drive model considered here is shown in Fig. 5.7. In this configuration, the lead screw is fixed in place and the nut rotates, causing it to translate along the lead screw together with other moving parts (i.e., motor, gearbox, payload, etc.). The equation of motion of this system is also given by (5.18).

5.4 Antibacklash Nut

As mentioned in Chap. 1, antibacklash nuts are commonly used to counter the effects of backlash and wear in a lead screw drive. An antibacklash nut is usually made of two parts that are connected together by a preloaded spring. Figure 5.8 shows a schematic model of a lead screw drive with a two-part nut. The spring k_n is preloaded such that a force $P = k_n\delta_n$ acts between the two halves of the nut. δ_n is the initial compression of the spring. Neglecting the mass of the nut, the Newton's second law gives

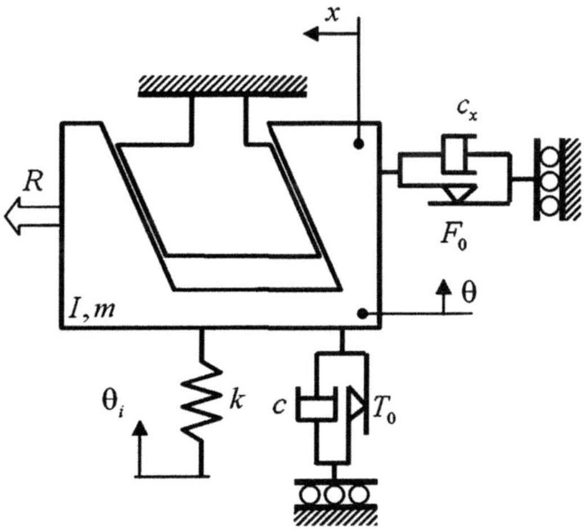

Fig. 5.6 Basic 1-DOF model with fixed nut

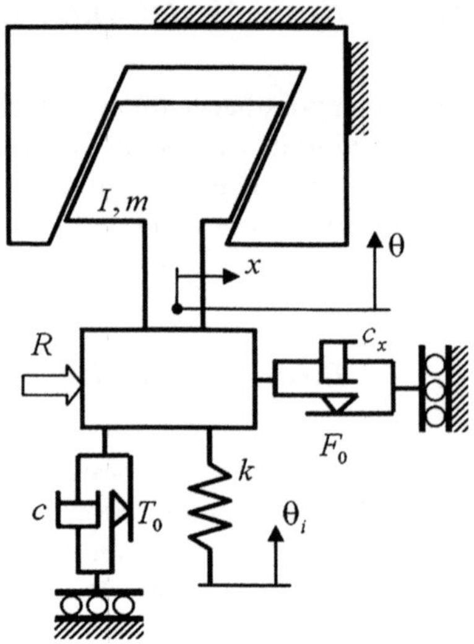

Fig. 5.7 Basic 1-DOF model with fixed lead screw

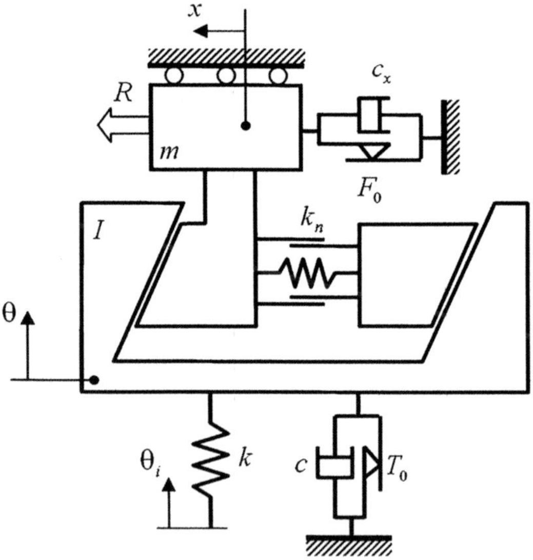

Fig. 5.8 Lead screw model with antibacklash nut

$$I\ddot{\theta} = k(\theta_{\rm i} - \theta) - c\dot{\theta} + r_{\rm m}\Big(N_1 \sin\lambda - \mu\,\mathrm{sgn}(\dot{\theta})N_1\cos\lambda\Big) + r_{\rm m}\Big(-N_2\sin\lambda - \mu\,\mathrm{sgn}(\dot{\theta})N_2\cos\lambda\Big) - T_0\,\mathrm{sgn}(\dot{\theta}), \tag{5.20}$$

$$m\ddot{x} = -c_x\dot{x} - N_1\cos\lambda - \mu\,\mathrm{sgn}(\dot{\theta})N_1\sin\lambda + R - F_0\,\mathrm{sgn}(\dot{x}) + P, \tag{5.21}$$

where

$$P = N_2\cos\lambda - \mu\,\mathrm{sgn}(\dot{\theta})N_2\sin\lambda, \tag{5.22}$$

where $N_1 > 0$ and $N_2 > 0$ are the thread contact forces corresponding to left and right parts of the nut, respectively.

Combining (5.20), (5.21), and (5.22) and using (2.3), gives

$$\begin{aligned}&\left(I - \tan\lambda\xi_1 m{r_\mathrm{m}}^2\right)\ddot{\theta} + k\theta + \left(c - {r_\mathrm{m}}^2\tan\lambda\xi c_x\right)\dot{\theta}\\&\quad = k\theta_\mathrm{i} - r_\mathrm{m}\xi_1\left(R - F_0\mathrm{sgn}(\dot{\theta})\right) - T_0\,\mathrm{sgn}(\dot{\theta}) - r_\mathrm{m}(\xi_1 + \xi_2)P,\end{aligned}$$

where

$$\xi_1 = \frac{\mu\,\mathrm{sgn}(\dot{\theta}) - \tan\lambda}{1 + \mu\,\mathrm{sgn}(\dot{\theta})\tan\lambda} \quad \text{and} \quad \xi_2 = \frac{\mu\,\mathrm{sgn}(\dot{\theta}) + \tan\lambda}{1 - \mu\,\mathrm{sgn}(\dot{\theta})\tan\lambda}.$$

Compared with (5.18), the term $-(\xi_1 + \xi_2)P$ is the additional resistive torque caused by the preloaded nut. The contact force N_1 (for left threads in Fig. 5.8) is found as

$$N_1 = \frac{\left(R - F_0\mathrm{sgn}(\dot{\theta})\right)I + mr_\mathrm{m}\tan\lambda\left[k(\theta - \theta_\mathrm{i}) + (c - {r_\mathrm{m}}^2\tan\lambda\xi c_x)\dot{\theta} + T_0\mathrm{sgn}(\dot{\theta})\right] + (I + \tan\lambda\xi_1 m{r_\mathrm{m}}^2)P}{\left(\cos\lambda + \mu\,\mathrm{sgn}(\dot{\theta})\sin\lambda\right)(I - \tan\lambda\xi_1 m{r_\mathrm{m}}^2)}.$$

Compared with (5.19), the contact force is increased due to the preload P.

Note that the above simplified formulation is valid as long as $N_1 > 0$. If this condition is violated (i.e., the left contact is lost) for the duration of such motion, the number of DOFs is increased to two. In such cases, which may be caused by large $-R > 0$, the dynamics of the system gets more complicated since the impact of the threads and repeated loss of contact should be considered.

5.5 Compliance in Lead Screw and Nut Threads

In Sect. 5.3, the lead screw and nut are modeled as a kinematic pair leading to an iterative equation for determining the sign of the contact force. The analysis may be greatly simplified by assuming some degree of compliance in the lead screw and/or nut threads. Figure 5.9 shows the same system as in Fig. 5.4(b) except for the

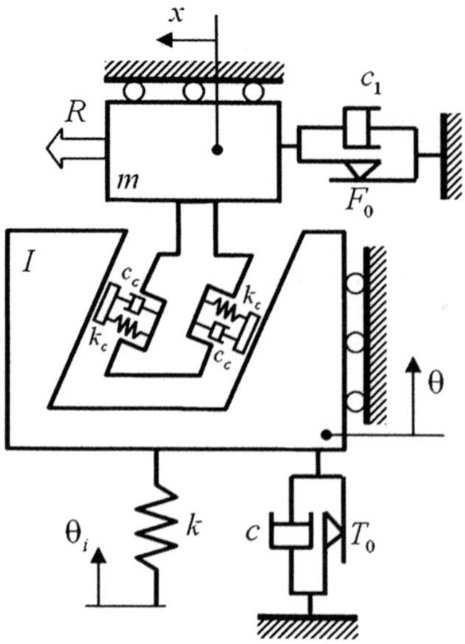

Fig. 5.9 2-DOF lead screw drive model including thread compliance

contact between threads which is now modeled by springs and dampers. With this change, the number of DOFs is increased to two.

Conforming to the sign convention defined in Fig. 2.7, the deflection (or interference) of threads can be calculated as

$$\delta = x\cos\lambda - r_{\mathrm{m}}\theta\sin\lambda. \tag{5.23}$$

The simplest way to approximate the contact force is by modeling the force–deflection relationship of the threads as that of linear springs and dampers. Thus

$$N = k_{\mathrm{c}}\delta + c_{\mathrm{c}}\dot{\delta}. \tag{5.24}$$

Substituting (5.24) into (5.16) and (5.17) and using (5.23) yields

$$\begin{aligned} I\ddot{\theta} = {} & k(\theta_{\mathrm{i}} - \theta) - c\dot{\theta} + r_{\mathrm{m}}k_{\mathrm{c}}(x\cos\lambda - r_{\mathrm{m}}\theta\sin\lambda)(\sin\lambda - \mu_{\mathrm{s}}\cos\lambda) \\ & + r_{\mathrm{m}}c_{\mathrm{c}}(\dot{x}\cos\lambda - r_{\mathrm{m}}\dot{\theta}\sin\lambda)(\sin\lambda - \mu_{\mathrm{s}}\cos\lambda) - T_0\,\mathrm{sgn}(\dot{\theta}), \end{aligned} \tag{5.25}$$

$$\begin{aligned} m\ddot{x} = {} & -c_x\dot{x} - k_{\mathrm{c}}(x\cos\lambda - r_{\mathrm{m}}\theta\sin\lambda)(\cos\lambda + \mu_{\mathrm{s}}\sin\lambda) \\ & - c_{\mathrm{c}}(\dot{x}\cos\lambda - r_{\mathrm{m}}\dot{\theta}\sin\lambda)(\cos\lambda + \mu_{\mathrm{s}}\sin\lambda) + R - F_0\,\mathrm{sgn}(\dot{x}), \end{aligned} \tag{5.26}$$

where μ_{s} is defined by (5.9).

Remark 5.1. For the model of this section, the relative sliding velocity is given by $v_{\mathrm{s}} = r_{\mathrm{m}}\dot{\theta}/\cos\lambda + \dot{\delta}\tan\lambda$ where δ is defined by (5.23). However, in practical situations where the lead angle (λ) is small, $v_{\mathrm{s}} \cong r_{\mathrm{m}}\dot{\theta}/\cos\lambda$. As a result, it is assumed that $\mathrm{sgn}(v_{\mathrm{s}}) = \mathrm{sgn}(\dot{\theta})$ which simplifies the equations of motion of the system. □

5.5.1 Backlash

Although we do not include backlash nonlinearity in our analysis of friction-induced vibration, it is worthwhile presenting a slightly modified version of the above model that enables one to include the effect of backlash, approximately.

Instead of (5.24), we may consider the following relationship for the contact force which is based on [114]:

$$N = \begin{cases} k_{\mathrm{c}}\left(\delta + \frac{\delta_{\mathrm{b}}}{2}\right)\left|\delta + \frac{\delta_{\mathrm{b}}}{2}\right|^{n} + c_{\mathrm{c}}\dot{\delta}\left|\delta + \frac{\delta_{\mathrm{b}}}{2}\right|^{n}, & \delta < -\frac{\delta_{\mathrm{b}}}{2}, \\ 0, & -\frac{\delta_{\mathrm{b}}}{2} \leq \delta \leq \frac{\delta_{\mathrm{b}}}{2}, \\ k_{\mathrm{c}}\left(\delta - \frac{\delta_{\mathrm{b}}}{2}\right)\left|\delta - \frac{\delta_{\mathrm{b}}}{2}\right|^{n} + c_{\mathrm{c}}\dot{\delta}\left|\delta - \frac{\delta_{\mathrm{b}}}{2}\right|^{n}, & \delta > \frac{\delta_{\mathrm{b}}}{2}, \end{cases} \tag{5.27}$$

where δ_{b} is the backlash measured perpendicular to the thread surface. Also $n \geq 1$ depend on the geometry of the contacting pair. Aside from the incorporation of backlash, the nonlinear damping relationship in (5.27) results in continuous contact force in transition between free motion and contact phases.

5.6 Axial Compliance in Lead Screw Supports

Another important source of flexibility in the system may be the compliance in the lead screw supports. To model this feature, as shown in Fig. 5.10, a spring k_1 and a damper c_1 are added to the basic model of Sect. 5.3, which allows the lead screw to move axially. For the sake of simplicity, in the remainder of this Chapter, the damping c_x is neglected.

Setting $c_x = 0$, (5.16) and (5.17) give force–acceleration relationships for the lead screw rotation and nut translation, respectively. Moreover, the lead screw translation DOF is governed by

$$m_1\ddot{x}_1 = -k_1 x_1 - c_1\dot{x}_1 + N\cos\lambda + F_{\mathrm{f}}\sin\lambda. \tag{5.28}$$

The kinematic relationship among θ, x, and x_1 is given as

$$x - x_1 = r_{\mathrm{m}}\tan\lambda\theta. \tag{5.29}$$

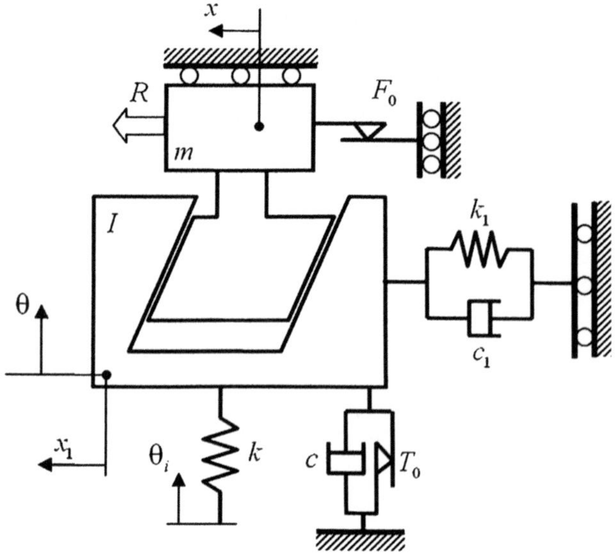

Fig. 5.10 2-DOF lead screw drive model including compliance in the supports

Eliminating N between (5.16) and (5.17) and also between (5.17) and (5.28) and using (5.29) and (2.4) yield

$$\left(I-\tan\lambda\xi m r_{\mathrm{m}}^{2}\right)\ddot{\theta}-r_{\mathrm{m}}\xi m\ddot{x}_{1}=k(\theta_{\mathrm{i}}-\theta)-c\dot{\theta}-r_{\mathrm{m}}\xi(R-F_{0}\,\mathrm{sgn}(\dot{x}))-T_{0}\,\mathrm{sgn}(\dot{\theta}), \quad (5.30)$$

$$(m_1+m)\ddot{x}_1+mr_{\mathrm{m}}\tan\lambda\ddot{\theta}=-k_1x_1-c_1\dot{x}_1+R-F_0\,\mathrm{sgn}(\dot{x}), \quad (5.31)$$

where ξ is given by (5.11).

The normal contact force is now calculated as

$$N=\frac{(I/m)(R-F_0\mathrm{sgn}(\dot{x}))+(I/m_1)k_1x_1+(I/m_1)c_1\dot{x}_1-r_{\mathrm{m}}\tan\lambda k(\theta_{\mathrm{i}}-\theta)+r_{\mathrm{m}}\tan\lambda\left(c\dot{\theta}+T_0\,\mathrm{sgn}(\dot{\theta})\right)}{(\cos\lambda+\mu_{\mathrm{s}}\sin\lambda)\left[((I/m)+(I/m_1))-r_{\mathrm{m}}^{2}\tan\lambda\xi\right]}. \quad (5.32)$$

5.6.1 Alternative Formulation

Following the formulation of Sect. 3.2, an alternative form of the equations of motion of the 2-DOF lead screw model with axial compliant lead screw support is derived in this section. Equations (5.16), (5.17), and (5.28) can be recast into (3.10) where $\mathbf{x}=\left[\theta \quad x \quad x_1\right]^{\mathrm{T}}$ and

$$\widehat{\mathbf{M}}(\mathbf{x}) = \mathbf{M} = \begin{bmatrix} I & 0 & 0 \\ 0 & m & 0 \\ 0 & 0 & m_1 \end{bmatrix}, \tag{5.33}$$

$$\mathbf{H}(\mathbf{x}, \dot{\mathbf{x}}) = \begin{bmatrix} k(\theta - \theta_i) + c\dot{\theta} + T_O \mathrm{sgn}\left(\dot{\theta}\right) \\ -R + F_O \mathrm{sgn}(\dot{x}) \\ k_1 x_1 + c_1 \dot{x}_1 \end{bmatrix}, \tag{5.34}$$

$$\mathbf{v}_n = \begin{bmatrix} r_m \sin\lambda & -\cos\lambda & \cos\lambda \end{bmatrix}^T, \tag{5.35}$$

$$\mathbf{v}_t = \begin{bmatrix} r_m \cos\lambda & \sin\lambda & -\sin\lambda \end{bmatrix}^T. \tag{5.36}$$

The constraint equation (5.29) is represented by (3.11) where

$$\mathbf{w}^T = \begin{bmatrix} r_m \tan\lambda & -1 & 1 \end{bmatrix}. \tag{5.37}$$

Furthermore, the transformation between the generalized coordinates and the reduced coordinates is given by (3.12) where $\mathbf{y} = [\theta \quad x_1]^T$ and

$$\mathbf{Q} = \begin{bmatrix} 1 & 0 \\ r_m \tan\lambda & 1 \\ 0 & 1 \end{bmatrix}.$$

The equation for the constraint normal force is given by (3.15). Upon substituting (5.33), (5.34), (5.35), (5.36), and (5.37) into (3.16) and (3.17), we get

$$A = (\cos\lambda + \mu_s \sin\lambda) I^{-1} \left[-r_m{}^2 \tan\lambda\,\xi + \left(\frac{I}{m} + \frac{I}{m_1} \right) \right], \tag{5.38}$$

$$b = \frac{r_m \tan\lambda}{I} \left[k(\theta - \theta_i) + c\dot{\theta} + T_O \mathrm{sgn}\left(\dot{\theta}\right) \right] - \frac{1}{m} [-R + F_O \mathrm{sgn}(\dot{x})] + \frac{1}{m_1} (k_1 x_1 + c_1 \dot{x}_1), \tag{5.39}$$

where ξ is given by (5.11). Note that, substituting (5.38) and (5.39) into (3.15) yields the same expression for the contact force as (5.32). The reduced order equations of motion is given by (3.18)

$$\tilde{\mathbf{M}}(\mathbf{y}, \dot{\mathbf{y}})\ddot{\mathbf{y}} + \tilde{\mathbf{H}}(\mathbf{y}, \dot{\mathbf{y}}) = 0, \tag{5.40}$$

where

$$\tilde{\mathbf{M}}(\mathbf{y}, \dot{\mathbf{y}}) = A \begin{bmatrix} I + m r_m{}^2 \tan^2\lambda & m r_m \tan\lambda \\ m r_m \tan\lambda & m + m_1 \end{bmatrix}, \tag{5.41}$$

$$\tilde{\mathbf{H}}(\mathbf{y},\dot{\mathbf{y}}) = A\begin{bmatrix} k(\theta-\theta_{\mathrm{i}})+c\dot{\theta}+T_0\,\mathrm{sgn}\left(\dot{\theta}\right)-r_{\mathrm{m}}\tan\lambda(R-F_0\,\mathrm{sgn}(\dot{x})) \\ k_1x_1+c_1\dot{x}_1-R+F_0\,\mathrm{sgn}(\dot{x}) \end{bmatrix} + \mu_{\mathrm{s}}b\begin{bmatrix} \dfrac{r_{\mathrm{m}}}{\cos\lambda} \\ 0 \end{bmatrix}.$$

Of course, (5.40) is equivalent to the system given by (5.30) and (5.31). In this representation, however, the possibility of Painlevé's paradox is clearly shown through the appearance of A, given by (5.38), in the equation of motion.

5.7 Compliance in Threads and Lead Screw Supports

By combining the two models presented in Sects. 5.5 and 5.6, a 3-DOF model of the lead screw drive is constructed as shown in Fig. 5.11. The equations of motion of this system are defined by (5.16), (5.17), and (5.28).

The only change is in the calculation of contact force N given by (5.24); the threads deflection, instead of (5.23), is calculated by

$$\delta = (x - x_1)\cos\lambda - r_{\mathrm{m}}\theta\sin\lambda.$$

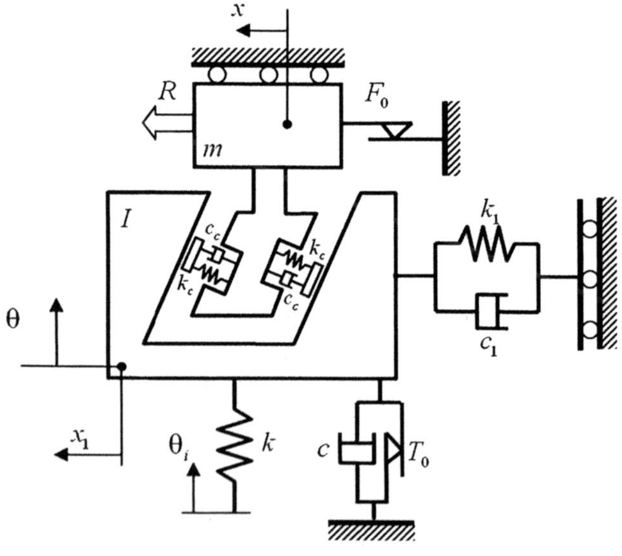

Fig. 5.11 3-DOF lead screw drive model including compliance in the supports and compliance in the lead screw and nut threads

5.8 Rotational Compliance of the Nut

Figure 5.12 shows a modified basic lead screw model where the nut has an additional rotational DOF (θ_2). Linear rotational spring and damper provide the rotational compliance.

Newton's second law for the lead screw rotation and nut translation yields relationships identical to (5.16) and (5.17), respectively. The kinematic constraint, instead of (2.3), is given by

$$x = r_{\mathrm{m}} \tan \lambda(\theta - \theta_2). \tag{5.42}$$

The rotational DOF of the nut is governed by the equation

$$I_2\ddot{\theta}_2 = -r_{\mathrm{m}}(N \sin \lambda - F_{\mathrm{f}} \cos \lambda) - k_2\theta_2 - c_2\dot{\theta}_2. \tag{5.43}$$

Eliminating N among (5.16), (5.17), and (5.43) yields the equations of motion for this 2-DOF lead screw drive model:

$$\begin{aligned}(I - \xi m r_{\mathrm{m}} \tan \lambda)\ddot{\theta} + \xi m r_{\mathrm{m}} \tan \lambda\ddot{\theta}_2 \\ = k(\theta_{\mathrm{i}} - \theta) - c\dot{\theta} - \xi(R - F_0\mathrm{sgn}(\dot{\theta} - \dot{\theta}_2)) - T_0\mathrm{sgn}(\dot{\theta})\end{aligned} \tag{5.44}$$

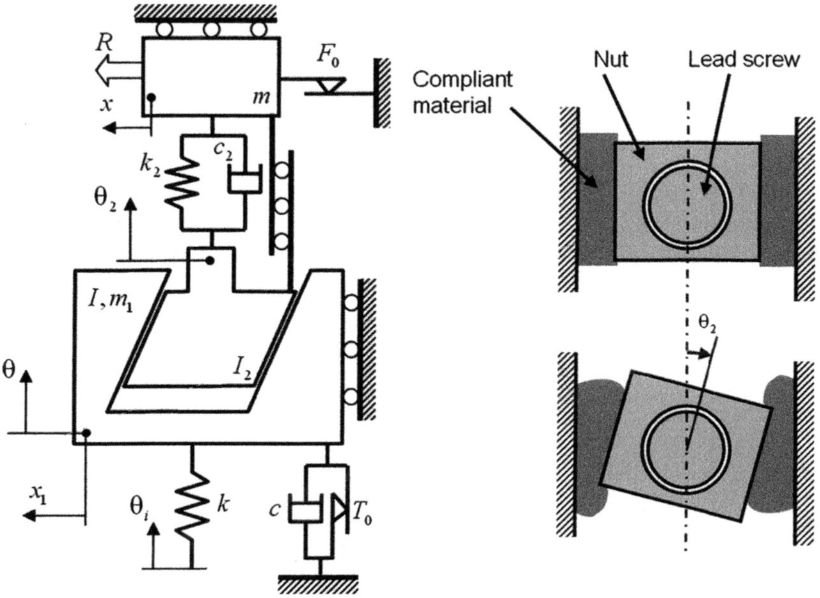

Fig. 5.12 2-DOF lead screw drive model with rotationally compliant nut

and

$$(I_2 - \xi m r_\mathrm{m} \tan\lambda)\ddot{\theta}_2 + \xi m r_\mathrm{m} \tan\lambda\ddot{\theta} = -k_2\theta_2 - c_2\dot{\theta}_2 + \xi(R - F_0 \operatorname{sgn}(\dot{\theta} - \dot{\theta}_2)), \tag{5.45}$$

where (5.42) was used and ξ is given by (5.11).

5.8.1 Alternative Formulation

Following the formulation of Sect. 3.2, an alternative form of the equations of motion of the 2-DOF lead screw model with rotational compliance of the nut is derived in this section. Equations (5.16), (5.17), and (5.43) can be recast into (3.10) where $\mathbf{x} = [\theta \quad x \quad \theta_2]^\mathrm{T}$ and

$$\widehat{\mathbf{M}}(\mathbf{x}) = \mathbf{M} = \begin{bmatrix} I & 0 & 0 \\ 0 & m & 0 \\ 0 & 0 & I_2 \end{bmatrix}, \tag{5.46}$$

$$\mathbf{H}(\mathbf{x}, \dot{\mathbf{x}}) = \begin{bmatrix} k(\theta - \theta_\mathrm{i}) + c\dot{\theta} + T_0 \operatorname{sgn}\left(\dot{\theta}\right) \\ -R + F_0 \operatorname{sgn}(\dot{x}) \\ k_2\theta_2 + c_2\dot{\theta}_2 \end{bmatrix}, \tag{5.47}$$

$$\mathbf{v}_\mathrm{n} = [r_\mathrm{m} \sin\lambda \quad -\cos\lambda \quad -r_\mathrm{m} \sin\lambda]^\mathrm{T}, \tag{5.48}$$

$$\mathbf{v}_\mathrm{t} = [r_\mathrm{m} \cos\lambda \quad \sin\lambda \quad -r_\mathrm{m} \cos\lambda]^\mathrm{T}, \tag{5.49}$$

The constraint equation (5.42) is represented by (3.11) where

$$\mathbf{w}^\mathrm{T} = [r_\mathrm{m} \tan\lambda \quad -1 \quad -r_\mathrm{m} \tan\lambda]. \tag{5.50}$$

Furthermore, the transformation between the generalized coordinates and reduced coordinates is given by (3.12) where $\mathbf{y} = [\theta \quad \theta_2]^\mathrm{T}$ and

$$\mathrm{Q} = \begin{bmatrix} 1 & 0 \\ r_\mathrm{m} \tan\lambda & -r_\mathrm{m} \tan\lambda \\ 0 & 1 \end{bmatrix}.$$

The equation for the constraint normal force is given by (3.15). Upon substituting (5.46), (5.34), (5.48), (5.49), and (5.50) into (3.16) and (3.17), we find

$$A = (\cos\lambda + \mu_\mathrm{s} \sin\lambda) \left[-r_\mathrm{m} \tan\lambda\xi \left(\frac{1}{I} + \frac{1}{I_2} \right) + \frac{1}{m} \right], \tag{5.51}$$

$$b = \frac{r_{\mathrm{m}} \tan\lambda}{I}\left[k(\theta - \theta_{\mathrm{i}}) + c\dot{\theta} + T_0 \operatorname{sgn}\left(\dot{\theta}\right)\right] - \frac{1}{m}[-R + F_0 \operatorname{sgn}(\dot{x})] - \frac{r_{\mathrm{m}} \tan\lambda}{I_2}\left(k_2\theta_2 + c_2\dot{\theta}_2\right), \tag{5.52}$$

where ξ is given by (5.11). The reduced order equations of motion is given by (3.18)

$$\tilde{\mathbf{M}}(\mathbf{y}, \dot{\mathbf{y}})\ddot{\mathbf{y}} + \tilde{\mathbf{H}}(\mathbf{y}, \dot{\mathbf{y}}) = 0, \tag{5.53}$$

where

$$\tilde{\mathrm{M}}(\mathbf{y}, \dot{\mathbf{y}}) = A\begin{bmatrix} I + r_{\mathrm{m}}{}^2\tan^2\lambda m & -r_{\mathrm{m}}{}^2\tan^2\lambda m \\ -r_{\mathrm{m}}{}^2\tan^2\lambda m & I_2 + r_{\mathrm{m}}{}^2\tan^2\lambda m \end{bmatrix},$$

$$\tilde{\mathrm{H}}(\mathbf{y}, \dot{\mathbf{y}}) = A\begin{bmatrix} k(\theta - \theta_{\mathrm{i}}) + c\dot{\theta} + T_0 \operatorname{sgn}\left(\dot{\theta}\right) - r_{\mathrm{m}} \tan\lambda[R - F_0 \operatorname{sgn}(\dot{x})] \\ -r_{\mathrm{m}} \tan\lambda[-R + F_0 \operatorname{sgn}(\dot{x})] + k_2\theta_2 + c_2\dot{\theta}_2 \end{bmatrix} + \mu_{\mathrm{s}} b \frac{r_{\mathrm{m}}}{\cos\lambda}\begin{bmatrix} 1 \\ -1 \end{bmatrix}.$$

The equation of motion (5.53) is equivalent to the system given by (5.44) and (5.45). In this representation, the possibility of Painlevé's paradox is clearly shown through the appearance of A, (5.51), in the equations.

5.9 Some Remarks Regarding the System Models

Depending on the configuration of an actual lead screw drive, one or more models presented in this chapter may be suitable to accurately capture the most prominent and/or relevant features of the system's dynamical behavior. This is certainly the case in the subsequent chapters. However, many other features are not included in this work. These features include:

- *Dependence of friction on position*: As the lead screw turns, the nut progresses along the lead screw threads creating the possibility of a position-dependent coefficient of friction. In this monograph, the friction model is assumed to be independent of position for the simplicity of mathematical modeling. From an experimental point of view, the identified friction (and other possible position-dependent parameters) may be considered as an averaged value over the working portion of the lead screw.
- *Nonlinearity*: The only nonlinear effect considered here comes from friction. However, many other sources of nonlinearity may exist that are excluded from the subsequent analyses to simplify the study of friction-induced vibrations in

lead screw drives. Most notable factors are the presence of nonlinearity in the contact forces of threads caused by deflection, the nonlinear torsional stiffness of the couplings, and the discontinuity due to backlash.

- *Torsional deflection of lead screw*: For a long and/or slender lead screw, the frequency of the first few torsional modes of vibration may be low enough to influence the system dynamics. Moreover, the winding/unwinding action of torsional deflections may affect the threads clearance and the overall load distribution causing further deviation from the models considered here. In this monograph, the lead screws are considered to be sufficiently stiff and modeled as rigid bodies.
- *Axial deflection of lead screw*: Similar to the previous point, this effect may influence the lead screw–nut interaction in two ways: by introducing new modes of vibration and by affecting the threads clearance and load distribution.
- *Lateral deflection of lead screw*: Three situations may lead to lateral deflection of the lead screw and additional modes of vibration: lateral loading, excessive axial loading causing buckling (a factor for long slender lead screws), and whirling (for very high rotation speeds).
- *Misalignment*: Design and/or assembly problems may lead to axial offset of the centerlines of lead screw and nut. The misalignment may also occur in the form of a skewed nut. In both of these cases, thread contact and load distribution may be affected severely.
- *Manufacturing issues*: Depending on the manufacturing method and the quality of the product, lead screws can suffer from lead error (particularly in longer designs). There may be external contaminants or surface defects on the lead screw or the nut. Although these and other similar issues may have significant impact on the function of a lead screw drive, they are excluded from this fundamental study on the friction-induced vibration.
- *Additional elements*: The study of lead screw drives, or any other mechanical system for that matter, can be augmented by other connected mechanical elements (e.g., a vibrating component on the moving part, additional DOF due to the flexibility of the moving part, external time-dependent forcing, etc.). These cases are outside the scope of this work and, depending on the problem they represent, may warrant a separate study.
- *Backlash*: Lead screw drives generally suffer from backlash. Here, backlash is not considered since the focus is on the effects of friction on power screws where the resisting load is considered to be constant and the system is considered to be moving with a constant input velocity. Backlash certainly will play a major role in the "positioning" applications of the lead screws.
- *Wear*: Throughout the operating life of a lead screw drive, wear causes changes to the contacting surfaces, thereby affecting the load distribution across the threads.

Chapter 6
Negative Damping Instability Mechanism

The conversion of rotary to translational motion in a lead screw system occurs at the meshing lead screw and nut threads. The contacting threads slide against each other creating a friction force opposing the direction of motion. The three main friction-induced instability mechanisms in dynamical systems were introduced in Chap. 4. In this chapter, the role of the velocity-dependent friction coefficient on the stability of lead screw systems is studied. We have seen in Sect. 4.1 that a decreasing coefficient of friction with relative sliding velocity can effectively act as a source of *negative damping* causing instabilities that lead to self-excited vibration. [1] The 1-DOF model of Sect. 5.3 is used in this chapter, which captures all the essential features of the lead screw system dynamics pertaining to the negative damping instability mechanism.

The equation of motion of the 1-DOF lead screw model is presented in Sect. 6.1. In Sect. 6.2, the eigenvalue analysis method is used to study the local stability of the steady-sliding state and the condition for the onset of the negative damping instability is found. This study is expanded in Sect. 6.3 using the method of first-order averaging. A complete picture of the stability properties of the system is obtained in this section. The results of the averaging analysis can also be used to predict the amplitude of vibrations when instability occurs and to study the effect of various system parameters on the steady-state vibrations. These results are important in the understanding of the role of friction-induced vibration on the generation of audible noise from a lead screw drive mechanism. A summary of results and conclusions is given in Sect. 6.4.

6.1 Equation of Motion

Neglecting F_0, T_0 and c_x for simplicity, (5.18) becomes

$$\Gamma\ddot{\theta} + k\theta + c\dot{\theta} = k\theta_{\mathrm{i}} - r_{\mathrm{m}}\xi R, \tag{6.1}$$

[1] In chapter 8, it will be shown that in systems with constant coefficient of friction, there are situations where a different instability mechanism can lead to negative damping instability.

O. Vahid-Araghi and F. Golnaraghi, *Friction-Induced Vibration in Lead Screw Drives*,
DOI 10.1007/978-1-4419-1752-2_6,

where

$$\Gamma = I - \tan\lambda\xi m r_{\mathrm{m}}^2 \tag{6.2}$$

and ξ is given by (5.11). Also, the velocity-dependent coefficient of friction is defined by (5.3). Let $z = \theta - \theta_{\mathrm{i}}$, $\dot{z} = \dot{\theta} - \Omega$, and $\ddot{z} = \ddot{\theta}$ where $\Omega = \mathrm{d}\theta_{\mathrm{i}}/\mathrm{d}t$ is a constant representing the input angular velocity. Substituting this change of variable into (6.1), gives

$$\Gamma\ddot{z} + c\dot{z} + kz = -c\Omega - r_{\mathrm{m}}\xi R. \tag{6.3}$$

At steady-sliding, we have $\ddot{z} = 0$, $\dot{z} = 0$, and $z = z_0$. Substituting these values in (6.3) yields

$$z_0 = -\frac{c\Omega + r_{\mathrm{m}}\xi_0 R}{k},$$

where

$$\xi_0 = \frac{\mu_0 \mathrm{sgn}(R\Omega) - \tan\lambda}{1 + \mu_0 \mathrm{sgn}(R\Omega)\tan\lambda} \tag{6.4}$$

and

$$\mu_0 = \mu_1 + \mu_2 \mathrm{e}^{-r_0|\Omega|} + \mu_3|\Omega|.$$

The change of variable $u = z - z_0$ converts (6.3) to

$$\Gamma\ddot{u} + c\dot{u} + ku = r_{\mathrm{m}}(\xi_0 - \xi)R. \tag{6.5}$$

Also, (5.3) becomes $\mu(\dot{u}) = \mu_1 + \mu_2 \mathrm{e}^{-r_0|\dot{u}+\Omega|} + \mu_3|\dot{u} + \Omega|$. Furthermore, the equation for the contact force, which is given by (5.19), is simplified to

$$N(u, \dot{u}) = \frac{\Gamma_0 R + m r_{\mathrm{m}} \tan\lambda(ku + c\dot{u})}{(\cos\lambda + \mu_{\mathrm{s}}\sin\lambda)\Gamma}, \tag{6.6}$$

where Γ_0 is found from (6.2) by replacing ξ with ξ_0 and the abbreviation (5.9) is now written as

$$\mu_{\mathrm{s}}(u, \dot{u}) = \mu(\dot{u})\mathrm{sgn}(N(u, \dot{u}))\mathrm{sgn}(\dot{u} + \Omega). \tag{6.7}$$

6.2 Local Stability of the Steady-Sliding State

The introduction of the new variables, $y_1 = u$ and $y_2 = \dot{u}$, converts (6.5) into a system of first-order differential equations. The Jacobian matrix of this system evaluated at the origin (i.e., steady-sliding equilibrium point) is found as

$$\mathbf{A} = \begin{bmatrix} 0 & 1 \\ -\dfrac{k}{\Gamma_0} & -\dfrac{c+\hat{c}}{\Gamma_0} \end{bmatrix}, \quad \Gamma_0 \neq 0,$$

where

$$\hat{c} = \frac{r_\mathrm{m}(1+\tan^2\lambda)|R|}{(1+\mu_0 \mathrm{sgn}(R\Omega)\tan\lambda)^2} d_\mu, \tag{6.8}$$

where d_μ is the gradient of the coefficient of friction curve vs. the relative velocity and is given by

$$d_\mu = -r_0\mu_2 \mathrm{e}^{-r_0|\Omega|} + \mu_3.$$

Note that $\hat{c}$ is the equivalent damping coefficient due to the velocity-dependent friction and it becomes negative if $d_\mu < 0$. The eigenvalues of the Jacobian matrix are

$$e_1, e_2 = -\frac{c+\hat{c}}{2\Gamma_0} \pm \frac{1}{2|\Gamma_0|}\sqrt{(c+\hat{c})^2 - 4k\Gamma_0}.$$

Assuming $\Gamma_0 > 0$,[2] the steady-sliding equilibrium point becomes unstable if

$$c + \hat{c} < 0. \tag{6.9}$$

The above instability threshold can be stated alternatively in terms of the applied axial force, R. The steady-sliding equilibrium point is unstable if

$$|R| > -c\frac{(1+\mu_0 \mathrm{sgn}(R\Omega)\tan\lambda)^2}{r_\mathrm{m}(1+\tan^2\lambda)d_\mu} \text{ and } d_\mu < 0. \tag{6.10}$$

The stable/unstable regions in the space of parameters R and d_μ are shown in Fig. 6.1. Expectedly, when negative friction damping is present ($d_\mu < 0$), there is a limiting value of axial force, beyond which the steady-sliding equilibrium point is unstable. This limit proportionally increases with the increase of the damping in the lead screw supports.

[2]Violation of this inequality also leads to instability, which is known as the "kinematic constraint instability mechanism." This instability mechanism is the subject of Chap. 8.

6.2.1 Numerical Examples

The parameter values used in the numerical examples presented here are given in Table 6.1. Most of these values are taken from the experimental study of Chap. 9.

For an axial force of $R = \pm 100$ N and input angular velocity of $\Omega = \pm 40$ rad/s, the critical damping coefficient is found according to (6.8) as $c_{cr} = \max\{0, -\hat{c}\}$. For the parameter values listed in Table 6.1, the critical damping value is found as

$$\text{If } \operatorname{sgn}(R\Omega) = +1 \text{ then } c_{cr} \approx 2.25 \times 10^{-4}$$

$$\text{If } \operatorname{sgn}(R\Omega) = -1 \text{ then } c_{cr} \approx 2.43 \times 10^{-4}$$

Figures 6.2 and 6.3 show the system trajectories for $c < c_{cr}$ and $c > c_{cr}$, respectively. In each simulation, the initial condition was $\mathbf{y}_0 = (0, -\Omega)$. As expected, for the damping level below (above) the critical value, the equilibrium point is unstable (stable). In the unstable cases, system trajectories are attracted to a limit cycle. Using the method of averaging, in the next section, the periodic solutions of the nonlinear equation of motion (limit cycles) are studied and the amplitude of steady-state vibrations is estimated.

It is interesting to note that, as shown in Fig. 6.2, in cases where force and angular velocity have the same sign (i.e., force assisting the motion), the displacement amplitude is considerably smaller than cases where the axial force resists the motion (i.e., $R\Omega < 0$).

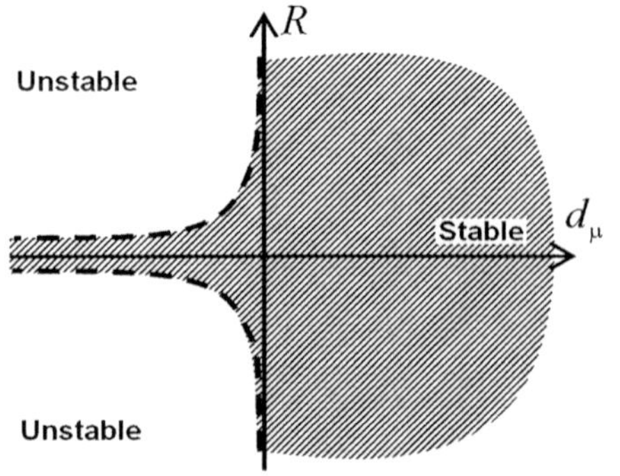

Fig. 6.1 Region of stability of the steady-sliding equilibrium point in the space of applied axial forces, R, and gradient of friction/velocity curve d_μ

Table 6.1 Parameter values used in the simulations

Parameter	Value	Parameter	Value
d_m	10.37 mm	μ_1	0.218
λ	5.57°	μ_2	0.0203
I	3.12×10^{-6} kg m^2	μ_3	-4.47×10^{-4} s/rad
k	1 N m/rad	r_0	0.38 rad/s
c	20×10^{-5} N m s/rad	R	± 100 N
m	3.8 kg	Ω	± 40 rad/s

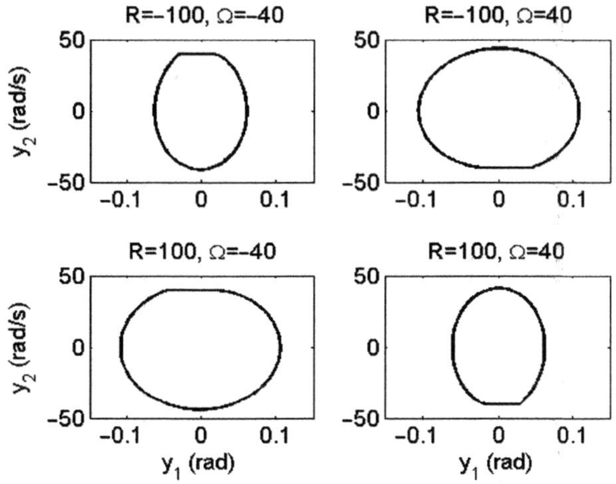

Fig. 6.2 System trajectories for $c = 2 \times 10^{-4} < c_{cr}$; unstable steady-sliding equilibrium point (0, 0)

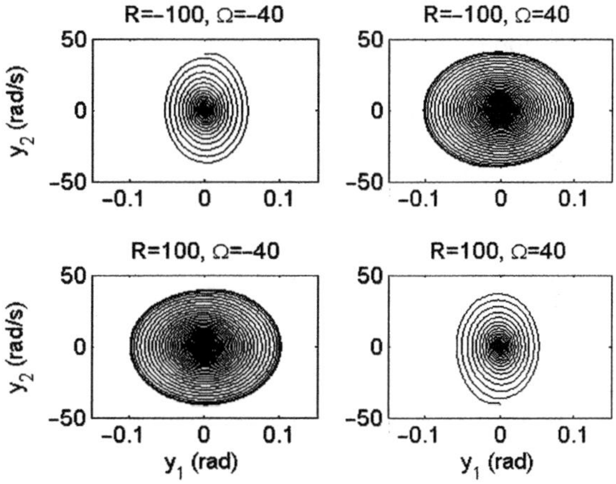

Fig. 6.3 System trajectories for $c = 3 \times 10^{-4} > c_{cr}$; stable steady-sliding equilibrium point (0, 0)

6.3 First-Order Averaging

The eigenvalue analysis of the previous section does not reveal any information regarding the behavior of the nonlinear system once instability occurs. The existence of periodic solutions (limit cycles), region of attraction of the stable trivial solution, and the effects of system parameters on these features as well as the size of the limit cycles (amplitude of steady-state vibrations) are important issues that are addressed in this section. The method of averaging[3] is used here to study the

[3]See Sect. 3.6.

behavior of the 1-DOF lead screw model as a *weakly nonlinear system*. For the lead screw equation of motion to be considered as a weakly nonlinear system, the friction and damping coefficients must be small. The relative smallness requirement of these parameters will be put into a more concrete setting later in the section.

Before performing averaging, (6.5) must be transformed to the standard form [56]. To that end, some simplifications are necessary. In the following sections, first the equation of motion is simplified and then converted into a nondimensionalized form. Next, a small parameter, ε, is introduced and the new dimensionless parameters are ordered to reach an approximate weakly nonlinear equation of motion accurate up to $O(\varepsilon)$.

6.3.1 Assumptions

As mentioned earlier, the current study is only concerned with the instability caused by negative damping. Thus, it is assumed that $\Gamma > 0$ for all $\dot{u}$.

From (6.7), it is easy to see that the equation of motion of the 1-DOF lead screw has a discontinuity whenever $\dot{\theta}$ crosses 0. To deal with this situation, the coefficient of friction is smoothed at zero relative velocity (i.e., $\dot{u} + \Omega = 0$) according to[4]

$$\mu(\dot{u}) = \left(\mu_1 + \mu_2 \mathrm{e}^{-r_0|\dot{u}+\Omega|}\right)\left(1 - \mathrm{e}^{-r_1|\dot{u}+\Omega|}\right) + \mu_3|\dot{u} + \Omega|, \tag{6.11}$$

where $r_1 > 0$ is a relatively large number. Substituting (6.11) into (6.7) yields

$$\mu_s(u,\dot{u}) = \left[\left(\mu_1 + \mu_2 \mathrm{e}^{-r_0|\dot{u}+\Omega|}\right)\left(1 - \mathrm{e}^{-r_1|\dot{u}+\Omega|}\right)\mathrm{sgn}(\dot{u}+\Omega) + \mu_3(\dot{u}+\Omega)\right]\mathrm{sgn}(N(u,\dot{u})). \tag{6.12}$$

It must be noted that, although (6.12) is discontinuous at $N(u,\dot{u}) = 0$, the differential equation of the system, given by (6.5), is continuous, since in its original form, given by (5.16) and (5.17), only the product μN appears.

From (6.6), we have

$$\mathrm{sgn}\, N = \mathrm{sgn}\left(\frac{\Gamma_0 R}{m r_{\mathrm{m}} \tan\lambda} + ku + c\dot{u}\right). \tag{6.13}$$

6.3.2 Equation of Motion in Standard Form

The first step toward transforming the equation of motion to a proper form for averaging is to nondimensionalize it. This is an important step to appropriately "order" each parameter according to its "size." Expanding (6.5), yields

[4] See the footnote on page 36.

$$\left(I - r_{\mathrm{m}}^2 \tan\lambda \frac{\mu_{\mathrm{s}} - \tan\lambda}{1 + \mu_{\mathrm{s}} \tan\lambda} m\right)\ddot{u} + c\dot{u} + ku$$
$$= r_{\mathrm{m}} \left(\frac{\mu_0 \mathrm{sgn}(R\Omega) - \tan\lambda}{1 + \mu_0 \mathrm{sgn}(R\Omega)\tan\lambda} - \frac{\mu_{\mathrm{s}} - \tan\lambda}{1 + \mu_{\mathrm{s}}\tan\lambda}\right) R. \tag{6.14}$$

Introduce the dimensionless time $\tau = \omega t$, where

$$\omega = \sqrt{\frac{k}{I}}. \tag{6.15}$$

The derivative with respect to τ is given as

$$\frac{\mathrm{d}(\cdot)}{\mathrm{d}t} = \omega \frac{\mathrm{d}(\cdot)}{\mathrm{d}\tau}. \tag{6.16}$$

Also, define nondimensional parameters

$$\widehat{m} = r_{\mathrm{m}}^2 \frac{m}{I} \tan\lambda, \tag{6.17}$$

$$\tilde{c} = \frac{c}{\sqrt{kI}}, \tag{6.18}$$

$$\widehat{R} = \frac{\omega}{|\Omega|} \frac{r_{\mathrm{m}}}{k} R, \quad \Omega \neq 0. \tag{6.19}$$

Using these new parameters, (6.14) is transformed to

$$\left(1 - \frac{\mu_{\mathrm{s}} - \tan\lambda}{1 + \mu_{\mathrm{s}}\tan\lambda}\widehat{m}\right) u'' + \tilde{c}u' + u = \frac{|\Omega|}{\omega}\left(\frac{\mu_0 \mathrm{sgn}(R\Omega) - \tan\lambda}{1 + \mu_0 \mathrm{sgn}(R\Omega)\tan\lambda} - \frac{\mu_{\mathrm{s}} - \tan\lambda}{1 + \mu_{\mathrm{s}}\tan\lambda}\right)\widehat{R}, \tag{6.20}$$

where prime denotes derivate with respect to τ. Now that the equation of motion is in its nondimensionalized form, based on physical insight, parameters are ordered using the small positive parameter ε. The new parameters,

$$\hat{\mu}_i = \frac{\mu_i}{\tan\lambda}, \quad i = 1, 2, 3,$$

$$\widehat{\mu}_0 = \frac{\mu_0}{\tan\lambda},$$

$$\widehat{c} = \frac{\tilde{c}}{\tan\lambda},$$

together with $\widehat{m}$ and $\widehat{R}$ are all assumed to be $O(1)$ with respect to ε where $\varepsilon = \tan\lambda$ is taken as the small parameter. Assuming, $|\Omega|/\omega = \rho\varepsilon$ where ρ is $O(1)$ and scaling u as

$$u = \varepsilon\rho v$$

gives

$$\left[1 - \varepsilon\Xi_1(v, v', \varepsilon)\widehat{m}\right]v'' + \varepsilon\widehat{c}v' + v = \varepsilon\widehat{R}[\Xi_0(\varepsilon) - \Xi_1(v', \varepsilon)], \tag{6.21}$$

where

$$\Xi_0(\varepsilon) = \frac{\mathrm{sgn}(R\Omega)\widehat{\mu}_0 - 1}{1 + \varepsilon^2\mathrm{sgn}(R\Omega)\widehat{\mu}_0}, \tag{6.22}$$

$$\Xi_1(v, v', \varepsilon) = \frac{\widehat{\mu}_s(v, v', \varepsilon) - 1}{1 + \varepsilon^2\widehat{\mu}_s(v', \varepsilon)}, \tag{6.23}$$

where the expression for the *signed* velocity-dependent coefficient of friction, $\widehat{\mu}_s(v', \varepsilon)$, in terms of the new dimensionless parameters, is

$$\begin{aligned}\widehat{\mu}_s(v, v', \varepsilon) = \Big[&\left(\hat{\mu}_1 + \hat{\mu}_2 \mathrm{e}^{-r_0||\Omega|v'+\Omega|}\right)\left(1 - \mathrm{e}^{-r_1||\Omega|v'+\Omega|}\right)\mathrm{sgn}(v' + \mathrm{sgn}(\Omega)) \\ &+ \hat{\mu}_3(|\Omega|v' + \Omega)\Big] \times \mathrm{sgn}\left(\frac{\widehat{R}}{\widehat{m}} - \varepsilon\Xi_0(\varepsilon)\widehat{R} + v + \varepsilon\widehat{c}v'\right).\end{aligned} \tag{6.24}$$

After rearranging, (6.21) becomes

$$v'' + v = \varepsilon f(v, v', \varepsilon), \tag{6.25}$$

where

$$f(v, v', \varepsilon) = -\left(1 - \varepsilon\Xi_1\widehat{m}\right)^{-1}\left[\widehat{c}v' + \widehat{m}\Xi_1 v + \widehat{R}(\Xi_1 - \Xi_0)\right]. \tag{6.26}$$

Remark 6.1. It is important to notice that, despite the presence of the two sign functions [i.e., $\mathrm{sgn}(v' + \mathrm{sgn}(\Omega))$ and $\mathrm{sgn}\left(\widehat{R}/\widehat{m} - \varepsilon\Xi_0(\varepsilon)\widehat{R} + v + \varepsilon\widehat{c}v'\right)$] in (6.24), $f(v, v', \varepsilon)$ is bounded and Lipschitz continuous with respect to its arguments for $(v, v', \varepsilon) \in D \times [0, \varepsilon_0]$, and D is any compact subset of $\mathbf{R}^2$ and $\varepsilon_0 > 0$ is some constant. To show this, we only need to investigate $f(v, v', \varepsilon)$ at instances where

$v' + \mathrm{sgn}(\Omega) = 0$ and $N = 0$ (which is equivalent to $\widehat{R}/\widehat{m} - \varepsilon\Xi_0(\varepsilon)\widehat{R} + v + \varepsilon\widehat{c}v' = 0$). For the first case, notice that $\widehat{\mu}_s(v, -\mathrm{sgn}(\Omega), \varepsilon) = 0$ and $\widehat{\mu}_s$ is continuous at $v' = -\mathrm{sgn}(\Omega)$, provided that $N \neq 0$. Furthermore,

$$\lim_{v' \to -1^+} \frac{\partial \widehat{\mu}_s(v, v', \varepsilon)}{\partial v'} = \lim_{v' \to -1^-} \frac{\partial \widehat{\mu}_s(v, v', \varepsilon)}{\partial v'} = (\hat{\mu}_1 + \hat{\mu}_2) r_1 \Omega \mathrm{sgn}(N), \ N \neq 0,$$

uniformly in ε. Also, $\partial\widehat{\mu}_s(v, v', \varepsilon)/\partial v = 0$ and $\partial\widehat{\mu}_s(v, v', \varepsilon)/\partial\varepsilon = 0$ for all (v, v') in the domain D and $N \neq 0$. For the second case, let $\delta(v, v', \varepsilon) = \widehat{R}/\widehat{m} - \varepsilon\Xi_0(\varepsilon)\widehat{R} + v + \varepsilon\widehat{c}v'$. Substituting this relationship into (6.26) gives $f(v, v', \varepsilon) = -\varepsilon^{-1}\left\{ \delta[1 - \varepsilon\Xi_1(v, v', \varepsilon)\widehat{m}]^{-1} - \left(\widehat{R}/\widehat{m} + v\right)\right\}$. Since $\Xi_1(v, v', \varepsilon)$ is bounded and continuous on $D \times [0, \varepsilon_0] - \{v, v', \varepsilon | \delta(v, v', \varepsilon) \neq 0\}$, and also since $1 - \varepsilon\Xi_1(v, v', \varepsilon)\widehat{m}$ is away from zero,[5] $f(v, v', \varepsilon)$ is continuous on $D \times [0, \varepsilon_0]$. Also it is easy to see that, $\lim_{\delta \to 0^+} \partial f/\partial v(v, v', \varepsilon)$, $\lim_{\delta \to 0^-} \partial f/\partial v(v, v', \varepsilon)$, $\lim_{\delta \to 0^+} \partial f/\partial v'(v, v', \varepsilon)$, $\lim_{\delta \to 0^-} \partial f/\partial v'(v, v', \varepsilon)$, $\lim_{\delta \to 0^+} \partial f/\partial\varepsilon(v, v', \varepsilon)$, and $\lim_{\delta \to 0^-} \partial f/\partial\varepsilon(v, v', \varepsilon)$ exist and are bounded, thus confirming the Lipschitz continuity of (6.26) with respect to its arguments. □

To transform (6.25) into the standard form, the following change of variables is used:

$$v = a\cos\varphi, \ v' = -a\sin\varphi. \tag{6.27}$$

This leads to

$$a' = -\varepsilon f(a\cos\varphi, -a\sin\varphi, \varepsilon)\sin\varphi, \tag{6.28}$$

$$\varphi' = 1 - \frac{\varepsilon}{a} f(a\cos\varphi, -a\sin\varphi, \varepsilon)\cos\varphi. \tag{6.29}$$

Remark 6.2. The change of variable (6.27) is only allowed in situations where the RHS of (6.29) remains bounded as a approaches 0 [58]. Here, this change of variables is allowed when R is away from 0, since after expanding (6.26) using power series, we get $f(a\cos\varphi, -a\sin\varphi, \varepsilon) = a\tilde{f}(a, \varphi, \varepsilon)$ for some bounded function $\tilde{f}$ for $0 \le a < a_0$ and for sufficiently small $a_0 < 1$ such that $N \neq 0$. □

Since φ' is away from 0, dividing (6.28) by (6.29) yields

$$\frac{\mathrm{d}a}{\mathrm{d}\varphi} = -\varepsilon \frac{f(a\cos\varphi, -a\sin\varphi, \varepsilon)\sin\varphi}{1 - (\varepsilon/a) f(a\cos\varphi, -a\sin\varphi, \varepsilon)\cos\varphi} \equiv \varepsilon g(\varphi, a, \varepsilon). \tag{6.30}$$

[5]This is the consequence of the initial assumption $\Gamma > 0$.

6.3.3 First-Order Averaging

In this section, the averaging method is applied to (6.30). To obtain the first-order averaged equations, the right-hand side of (6.30) must be averaged over a period (i.e., $T = 2\pi$) while keeping a as constant[6]. This gives

$$\begin{aligned} a' &= \frac{\varepsilon}{2\pi}\int_0^{2\pi} g(\varphi, a, 0)\,\mathrm{d}\varphi, \\ &= -\frac{\varepsilon}{2\pi}\int_0^{2\pi} f(a\cos\varphi, -a\sin\varphi, 0)\sin\varphi\,\mathrm{d}\varphi. \end{aligned} \tag{6.31}$$

Theorems presented in Sect. 3.6 establish the error estimate of the averaged system, (6.31), with respect to the original differential equation, (6.30). Substituting (6.27) into (6.26) and then substituting the result into (6.31) gives

$$\begin{aligned} a' = &-\frac{\varepsilon}{2\pi}\int_0^{2\pi}\left(\widehat{c}a\sin^2\varphi + \widehat{m}a\sin\varphi\cos\varphi + \widehat{\mu}_0\widehat{R}\sin\varphi\right)\mathrm{d}\varphi \\ &+\frac{\varepsilon}{2\pi}\int_0^{2\pi}\left(\widehat{m}a\sin\varphi\cos\varphi + \widehat{R}\sin\varphi\right)\widehat{\mu}_\mathrm{s}\mathrm{d}\varphi. \end{aligned} \tag{6.32}$$

After carrying out the integration of the first term, (6.32) becomes

$$a' = -\varepsilon\frac{\widehat{c}a}{2} + \frac{\varepsilon}{2\pi}\int_0^{2\pi}\left(\widehat{m}a\sin\varphi\cos\varphi + \widehat{R}\sin\varphi\right)\widehat{\mu}_\mathrm{s}(\varphi, a)\mathrm{d}\varphi, \tag{6.33}$$

where

$$\begin{aligned} \widehat{\mu}_\mathrm{s}(\varphi, a) = \Big[&\left(\hat{\mu}_1 + \hat{\mu}_2\mathrm{e}^{-r_0|\Omega - |\Omega|a\sin\varphi|}\right)\left(1 - \mathrm{e}^{-r_1|\Omega - |\Omega|a\sin\varphi|}\right)\mathrm{sgn}(\Omega - |\Omega|a\sin\varphi) \\ &+\hat{\mu}_3(\Omega - |\Omega|a\sin\varphi)\Big] \times \mathrm{sgn}\left(\frac{\widehat{R}}{\widehat{m}} + a\cos\varphi\right). \end{aligned}$$

The averaged differential equation given by (6.33) is too complicated to be approached analytically. Limiting our study to the situations where $\Omega > 0$ and also where $R > 0$ is large enough such that N remains positive over the domain of interest, $\widehat{\mu}_\mathrm{s}$ simplifies to

$$\begin{aligned} \widehat{\mu}_\mathrm{s}(\varphi, a) = &\left[\left(\hat{\mu}_1 + \hat{\mu}_2\mathrm{e}^{-r_0\Omega|1 - a\sin\varphi|}\right)\left(1 - \mathrm{e}^{-r_1\Omega|1 - a\sin\varphi|}\right) + \hat{\mu}_3\Omega(1 - a\sin\varphi)\right] \\ &\times \mathrm{sgn}(1 - a\sin\varphi). \end{aligned} \tag{6.34}$$

[6]For simplicity of notation, from this point on, prime denotes differentiation with respect to φ.

Substituting (6.34) into (6.33) and simplifying yields

$$a' = -\varepsilon\frac{\widehat{c}a}{2} + \frac{\varepsilon}{2\pi}\widehat{R}\int_0^{2\pi} \sin\varphi\, \widehat{\mu}_s(\phi, a)\mathrm{d}\varphi. \tag{6.35}$$

In addition to the assumption of $N > 0$, if the maximum amplitude is limited to one, i.e., $0 \le a \le 1$, (6.34) further simplifies to

$$\widehat{\mu}_s(\varphi, a) = \left(\widehat{\mu}_1 + \widehat{\mu}_2 \mathrm{e}^{r_0\Omega a \sin\varphi}\right)\left(1 - r_2 \mathrm{e}^{r_1\Omega a \sin\varphi}\right) + \widehat{\mu}_3(1 - a\sin\varphi), \tag{6.36}$$

where $r_2 = \mathrm{e}^{-r_1\omega}$ and also

$$\widehat{\mu}_1 = \hat{\mu}_1, \tag{6.37}$$

$$\widehat{\mu}_2 = \hat{\mu}_2 \mathrm{e}^{-r_0\Omega}, \tag{6.38}$$

$$\widehat{\mu}_3 = \hat{\mu}_3\Omega. \tag{6.39}$$

Substituting (6.36) into (6.35) and simplifying give

$$\begin{aligned} a' &= -\varepsilon\frac{\left(\widehat{c} + \widehat{\mu}_3\widehat{R}\right)a}{2} + \frac{\varepsilon}{2\pi}\widehat{\mu}_2\widehat{R}\int_0^{2\pi} \sin\varphi \mathrm{e}^{r_0\Omega a \sin\varphi}\mathrm{d}\varphi \\ &- \frac{\varepsilon}{2\pi} r_2\widehat{\mu}_1\widehat{R}\int_0^{2\pi} \sin\varphi \mathrm{e}^{r_1\Omega a \sin\varphi}\mathrm{d}\varphi - \frac{\varepsilon}{2\pi} r_2\widehat{\mu}_2\widehat{R}\int_0^{2\pi} \sin\varphi \mathrm{e}^{(r_0+r_1)\Omega a \sin\varphi}\mathrm{d}\varphi. \end{aligned} \tag{6.40}$$

Carrying out the rest of the integrations, one finds

$$a' = -\varepsilon\frac{\widehat{c} + \widehat{\mu}_3\widehat{R}}{2}a + \varepsilon\widehat{\mu}_2\widehat{R}\Lambda(r_0\Omega a) - \varepsilon r_2\widehat{\mu}_1\widehat{R}\Lambda(r_1\Omega a) - \varepsilon r_2\widehat{\mu}_2\widehat{R}\Lambda((r_0 + r_1)\Omega a), \tag{6.41}$$

where

$$\Lambda(\zeta) = \sum_{n=1}^{\infty} \frac{n}{2^{2n-1}(n!)^2}\zeta^{2n-1}. \tag{6.42}$$

In the next section, steady-state solutions of (6.41) are studied.

Remark 6.3. In cases where stable (unstable) nontrivial solutions exist and $a = a^* \le 1$, the above averaging process guarantees that the original system, (6.20), has stable (unstable) limit cycle in an $O(\varepsilon)$ neighborhood of the circle $r = \Omega a^*/\omega$, with a period $O(\varepsilon)$ close to 2π for sufficiently small $\varepsilon > 0$[58]. □

6.3.4 Steady-Sliding Equilibrium Point

Substituting (6.42) into (6.41) and rearranging

$$a' = \varepsilon\left(b_0 a + \sum_{n=1}^{\infty} b_n a^{2n+1}\right), \tag{6.43}$$

where

$$b_0 = -\frac{\widehat{c}}{2} + \frac{\widehat{R}}{2}\left(-\widehat{\mu}_3 + \widehat{\mu}_2 \Omega r_0 - r_2 \widehat{\mu}_1 \Omega r_1 - r_2 \widehat{\mu}_2 \Omega (r_0 + r_1)\right) \tag{6.44}$$

and

$$b_n = \frac{\Omega^{2n+1}\widehat{R}}{2^{2n+1} n!(n+1)!}\left(\widehat{\mu}_2 {r_0}^{2n+1} - r_2 \widehat{\mu}_1 {r_1}^{2n+1} - r_2 \widehat{\mu}_2 (r_0 + r_1)^{2n+1}\right). \tag{6.45}$$

It is obvious that $a = 0$ is the trivial solution. To determine its stability, $\mathrm{d}a'/\mathrm{d}a$ is derived and evaluated at $a = 0$. From (6.43)

$$\left.\frac{\mathrm{d}a'}{\mathrm{d}a}\right|_{a=0} = \varepsilon b_0. \tag{6.46}$$

From (6.46), the condition for the stability of the trivial solution is $b_0 < 0$. From (6.44) and by substituting the original system parameters, we have

$$\varepsilon b_0 = \frac{1}{2}(-c + c_{\mathrm{cr}}), \tag{6.47}$$

where

$$c_{\mathrm{cr}} = -r_{\mathrm{m}} R \left.\frac{\partial \mu}{\partial \dot{u}}\right|_{\dot{u}=0}. \tag{6.48}$$

From (6.47), the condition for the stability of the steady-sliding state becomes

$$c > c_{\mathrm{cr}}. \tag{6.49}$$

It is interesting to note that, (6.48) is accurate to $O(\varepsilon^2)$ when compared with what was found from linear eigenvalue analysis, i.e., (6.8):

$$c_{\mathrm{cr}} = -r_{\mathrm{m}} R \frac{1 + \tan^2\lambda}{(1 + \mu_0 \tan\lambda)^2} \left.\frac{\partial \mu}{\partial y_2}\right|_{y_2=0}, \quad R > 0,\ \Omega > 0. \tag{6.50}$$

Unfortunately, the other possible solutions (i.e., stable or unstable limit cycles) can only be found numerically due to the complexity of the averaged equations. However, some important insights can be gained by examining (6.43).

6.3.5 Nontrivial Equilibrium Points

To investigate the existence of the nontrivial equilibrium points (i.e., $a > 0$), we divide (6.43) by a and set the result to 0 which gives

$$b_0 + \sum_{n=1}^{\infty} b_n a^{2n} = 0, \tag{6.51}$$

which is a polynomial equation in a^2. Depending on the value of μ_2 and μ_3, four cases can be identified for the variation of the coefficient of friction with velocity. These cases are presented next.

Case 1. $\mu_2 = 0$ and $\mu_3 \geq 0$.

As shown in Fig. 6.4a, the gradient of the coefficient of friction function is positive for every $\Omega > 0$, i.e., $d\widehat{\mu}_s/dv'\big|_{v'=0} > 0$ thus from (6.47); $b_0 < 0$ and for any $c \geq 0$, and the trivial solution is stable. To investigate the possibility of nontrivial solutions, (6.40) is examined; setting $\mu_2 = 0$ and rearranging gives

$$a' = -\varepsilon\left(\frac{\widehat{c} + \widehat{\mu}_3\widehat{R}}{2}\right)a - \frac{\varepsilon}{2\pi} r_2\widehat{\mu}_1\widehat{R}\int_0^{2\pi} \sin\varphi e^{r_1\Omega a\sin\varphi} d\varphi. \tag{6.52}$$

From (6.42), we know that the definite integral in (6.52) is positive for $a > 0$ and 0 for $a = 0$. Since the first term is linear in a with a negative slope, one concludes that $a'(a) < 0$ for $0 < a \leq 1$, and there are no other nontrivial equilibrium points. A typical plot of amplitude equation, (6.40), for this case is shown in Fig. 6.4b.

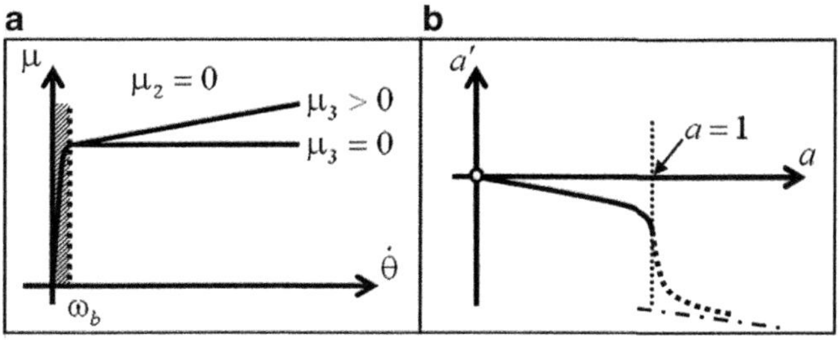

Fig. 6.4 Case 1 – $\mu_2 = 0$ and $\mu_3 \geq 0$

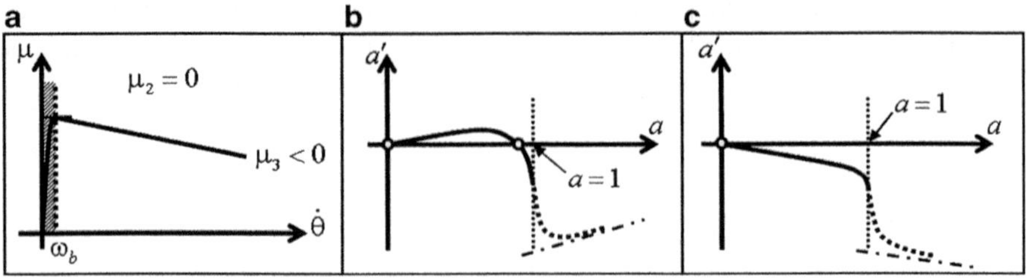

Fig. 6.5 Case 2 – $\mu_2 = 0$ and $\mu_3 < 0$

Case 2. $\mu_2 = 0$ and $\mu_3 < 0$.

A schematic plot of the variation of the coefficient of friction for this case is shown in Fig. 6.5a. First, we notice that for small velocities satisfying $0 < \Omega < \omega_b$ where $\omega_b = -(1/r_1)\ln(-(\hat{\mu}_3/\hat{\mu}_1 r_1))$ (i.e., $\dot{\theta}$ maximum of the friction curve), (6.49) is satisfied for any $c \geq 0$. Thus, the trivial solution is stable and $b_0 < 0$. Moreover, since all the coefficients of (6.43) are nonpositive, no other solution exists.

For $\Omega > \omega_b$, from (6.49) the trivial solution is stable if $\hat{c} > \hat{R}(-\hat{\mu}_3 - r_2\hat{\mu}_1\Omega r_1) > 0$ and it is unstable otherwise. In the case of stable trivial equilibrium point, once again all the coefficients of (6.51) are nonpositive, which implies that no other solutions are possible.

If (6.49) is not satisfied, b_0 is positive while the rest of the coefficients, b_n [given by (6.45)], remain less than or equal to 0;

$$b_n = -\hat{R} r_2 \hat{\mu}_1 \frac{\Omega^{2n+1} r_1{}^{2n+1}}{2^{2n+1} n!(n+1)!}.$$

According to Descartes' Rule of Signs [68], (6.51) has a positive solution which corresponds to a stable periodic solution of the original system. The schematic plots of the amplitude equation, (6.41), when trivial equilibrium point is unstable and when it is stable are given in Fig. 6.5b, c, respectively.

Case 3. $\mu_2 > 0$ and $\mu_3 \leq 0$.

A schematic plot of the variation of the coefficient of friction for this case is shown in Fig. 6.6a. The friction curves reach the maximum at $\Omega = \omega_b$ where ω_b is the solution of $-\hat{\mu}_3 + \hat{\mu}_2\Omega r_0 - r_2\hat{\mu}_1\Omega r_1 - r_2\hat{\mu}_2\Omega(r_0 + r_1) = 0$. Similar to the previous case, at low velocities, i.e., $0 < \Omega < \omega_b$,

$$-\hat{\mu}_3 + \hat{\mu}_2\Omega r_0 - r_2\hat{\mu}_1\Omega r_1 - r_2\hat{\mu}_2\Omega(r_0 + r_1) < 0 \tag{6.53}$$

and (6.49) is satisfied for any $c \geq 0$ and $b_0 < 0$. Thus, the trivial solution is stable. Multiplying (6.53) by $r_0{}^{2n}$ yields

$$-r_0{}^{2n}\hat{\mu}_3 - r_0{}^{2n} r_1 r_2 \hat{\mu}_1 \Omega - r_0{}^{2n}(r_0 + r_1) r_2 \hat{\mu}_2 \Omega + r_0{}^{2n+1}\hat{\mu}_2 \Omega < 0. \tag{6.54}$$

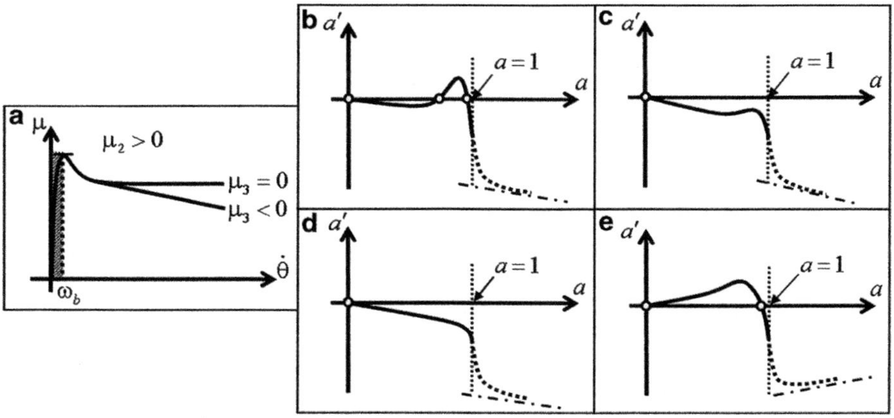

Fig. 6.6 Case 3 – $\mu_2 > 0$ and $\mu_3 \le 0$

Since ${r_0}^{2n} < (r_0 + r_1)^{2n}$, ${r_0}^{2n} < {r_1}^{2n}$, and $\widehat{\mu}_3 \le 0$, from (6.54) one gets

$$-r_2{r_1}^{2n+1}\widehat{\mu}_1 - (r_0 + r_1)^{2n+1} r_2\widehat{\mu}_2 + \widehat{\mu}_2{r_0}^{2n+1} < 0. \tag{6.55}$$

Consequently, from (6.55) it is obvious that $b_n < 0$. Thus, (6.51) has no other solutions.

When $\Omega > \omega_b$, the trivial solution is stable if

$$\widehat{c} > \widehat{R}\Big(-\widehat{\mu}_3 + \widehat{\mu}_2\Omega r_0 - r_2\widehat{\mu}_1\Omega r_1 - r_2\widehat{\mu}_2\Omega(r_0 + r_1)\Big). \tag{6.56}$$

If (6.56) holds $b_0 < 0$ otherwise $b_0 > 0$. The sign of b_n from (6.45) is the same as the sign of

$$\chi_n = \widehat{\mu}_2{r_0}^{2n+1} - r_2\widehat{\mu}_1{r_1}^{2n+1} - r_2\widehat{\mu}_2(r_0 + r_1)^{2n+1}. \tag{6.57}$$

Substituting (6.37), (6.38), and (6.39) into (6.57) and rearranging yields

$$\chi_n = \hat{\mu}_2\Big(\mathrm{e}^{-r_0\Omega}{r_0}^{2n+1} - \mathrm{e}^{-(r_0+r_1)\Omega}(r_0 + r_1)^{2n+1}\Big) - \mathrm{e}^{-r_1\Omega}\hat{\mu}_1{r_1}^{2n+1}.$$

Obviously, if $\mathrm{e}^{-r_0\omega}{r_0}^{2n+1} - \mathrm{e}^{-(r_0+r_1)\omega}(r_0 + r_1)^{2n+1} < 0$ then $\chi_n < 0$ regardless of the value of $\hat{\mu}_1$ and $\hat{\mu}_2$. Define

$$v = \frac{1}{2}\left[\frac{r_1\Omega}{\ln(r_0 + r_1) - \ln(r_0)} - 1\right].$$

If $v > 1$ then for all $1 \le n < v$ we have $\mathrm{e}^{-r_0\Omega}{r_0}^{2n+1} - \mathrm{e}^{-(r_0+r_1)\Omega}(r_0 + r_1)^{2n+1} > 0$. Consequently, there exists $\widehat{\mu}_2^*$ dependent on $\hat{\mu}_1$ (and other parameters) such that for

$\widehat{\mu}_2 > \widehat{\mu}_2^*$, χ_n is positive for $1 \leq n < v$. Of course, in this case for $n \geq v$, we have $\chi_n < 0$. Consequently, there is one sign change in b_n's, $n \geq 1$. On the other hand, if $v \not> 1$ then $\chi_n < 0$ for all $n \geq 1$ and there is no sign change in b_n's.

Depending on the system parameters values, one of the following three scenarios may happen:

Scenario 1. If (6.56) holds and $v > 1$ and $\widehat{\mu}_2 > \widehat{\mu}_2^*$, then the polynomial equation given by (6.51) has two sign changes in its coefficients. According to Descartes' Rule of Signs, (6.51) can have either two or zero positive roots. Typical plots of a' as a function of a are shown in Fig. 6.6b, c. The value of the damping, c, determines which case describes the system's behavior.

Scenario 2. If (6.56) holds and $v > 1$ and $\widehat{\mu}_2 < \widehat{\mu}_2^*$, then the polynomial equation given by (6.51) has no sign changes. As a result, the trivial solution is stable and no other solution exists. A typical plot of a' as a function of a is similar to the one shown in Fig. 6.6d.

Scenario 3. If (6.56) does not hold, then the polynomial equation given by (6.51) has only one sign change. As a result, there is only one nontrivial solution. In this case, the trivial solution is unstable and there is a stable limit cycle. A typical plot of a' as a function of a is shown in Fig. 6.6e.

Case 4. $\mu_2 > 0$ *and* $\mu_3 > 0$.

In this case, the same scenarios as in Case 3 are possible. Comparing to Case 3 with $\mu_3 = 0$, in order to consider the case of positive μ_3, the negative term $-\varepsilon(\widehat{\mu}_3\widehat{Ra}/2) < 0$ must be added to (6.43). This addition decreases $a'(a)$ throughout $0 < a \leq 1$ and may convert situation shown in Fig. 6.6b to the one shown in Fig. 6.6c. Also the situation shown in Fig. 6.6e may be converted to the situation shown in either Fig. 6.6b or c.

Remark 6.4. Cases where $\mu_2 < 0$ are not considered since the term $\mu_2 e^{-r_0\Omega}$ is added to the friction model only to emulate the Stribeck effect (i.e., decreasing the coefficient of friction with increasing relative velocity at low velocities). □

Remark 6.5. Because of the smoothing of the coefficient of friction, regardless of the value of μ_2 and μ_3, the steady-sliding state is stable for very low input angular velocities (i.e., $0 < \Omega < \omega_b$). □

Remark 6.6. The preceding analysis shows that there exists a stabilizing damping level, c_{st}, such that the system equilibrium point is stable for $c \geq c_{st}$ and no limit cycles exist. $c_{st} = 0$ in Case 1 and Cases 3 and 4/Scenario 2. $c_{st} = c_{cr}$ in Case 2 and Cases 3 and 4/Scenario 3, where c_{cr} is given by (6.48). $c_{st} > c_{cr}$ in Cases 3 and 4/Scenario 1. □

Remark 6.7. In cases where there is a stable nontrivial equilibrium point, $a = a^*$, there may be parameter values such that the amplitude of vibration is greater than 1 which violates the assumptions leading to (6.40). In these cases, (6.35) or even (6.33) may be used to calculate the averaged amplitude of vibrations numerically with $O(\varepsilon)$ accuracy. □

From the above arguments, it can be deduced that depending on the system parameters one of the following three cases defines the dynamic behavior of the averaged system:

1. The trivial solution is stable and no other solution exists.
2. The trivial solution is stable and it is surrounded by an unstable limit cycle, which defines the region of attraction of the trivial solution. The unstable limit cycle is inside a stable limit cycle.
3. The trivial solution is unstable and it is surrounded by a stable limit cycle.

6.3.6 *Numerical Simulation Results: Part 1*

In the examples presented here, unless otherwise specified, the system's parameter values are those listed in Table 6.2.

For the parameter values and the initial conditions selected, all simulation results satisfy $v' \geq -1$ and $N > 0$ conditions. As a result, the simplified averaged system equation given by (6.40) or (6.43) is used. Computationally, it is much more efficient to use (6.40) instead of the infinite sum of (6.43).

Figures 6.7 and 6.8 show comparisons between the numerical integration of the approximate (truncated) equation of motion given by

$$v'' + v = \varepsilon f(v, v', 0)$$

and the equilibrium points of the averaged amplitude equation, (6.40), for the two values of the lead screw damping; $c = 2 \times 10^{-4} < c_{\text{cr}}$ and $c = 3 \times 10^{-4} > c_{\text{cr}}$, respectively. Note that in these figures, both amplitudes are scaled by Ω/ω to reflect the physical system's vibration levels. Results show very accurate prediction of the steady-state amplitude of vibration by the first-order averaging method. However, when compared with the original (untruncated) equation of motion, (6.25), the averaging results have some differences as shown in Figs. 6.9 and 6.10. This deviation is caused by the effects of the omitted higher-order terms in the first-order averaging process.

Table 6.2 Parameter values used in the simulations

Parameter	Value	Parameter	Value
d_{m}	10.37 mm	μ_1	0.218
λ	5.57°	μ_2	0.0203
I	3.12×10^{-6} kg m^2	μ_3	-4.47×10^{-4} s/rad
k	1 N m/rad	r_0	0.38 rad/s
c	20×10^{-5} N m s/rad	r_1	2 rad/s
m	1 kg	Ω	40 rad/s
R	100 N		

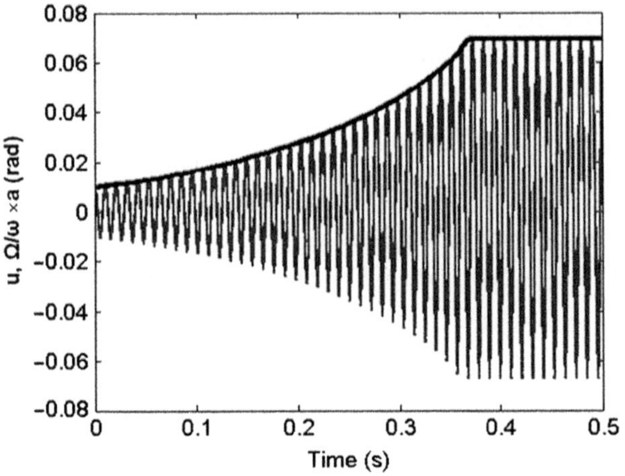

Fig. 6.7 First-order averaging results. $c = 2 \times 10^{-4}$; *grey*: truncated equation of motion; *black*: amplitude of vibration from first-order averaging

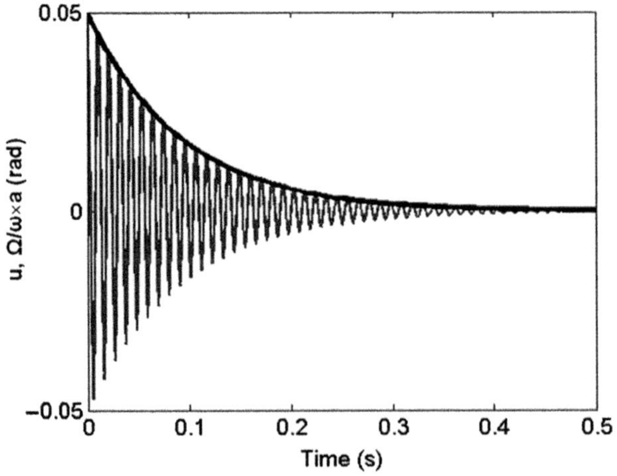

Fig. 6.8 First-order averaging results. $c = 3 \times 10^{-4}$; *grey*: truncated equation of motion; *black*: amplitude of vibration from first-order averaging

It must be noted that, the steady-sliding vibration amplitude in Figs.6.9 and 6.10 are still predicted very accurately by the averaged equation for the parameter values given in Table 6.2.

Figure 6.11 shows the bifurcation diagram of the amplitude equation, (6.40), where the damping coefficient, c, is taken as the control parameter. The trivial solution (i.e., the equilibrium point of the original system) undergoes a subcritical pitchfork bifurcation [69] at approximately $c_{cr} = 2.32 \times 10^{-4}\,\mathrm{Nm\,s/rad}$. This value agrees with (6.50). It can be shown that this bifurcation corresponds to a

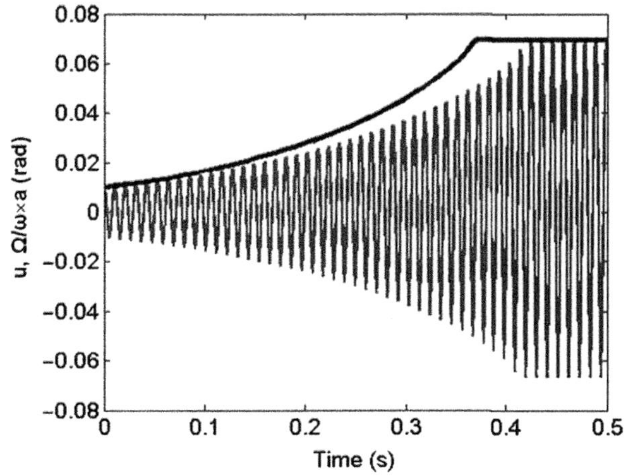

Fig. 6.9 First-order averaging results. $c = 2 \times 10^{-4}$; *grey*: original equation of motion; *black*: amplitude of vibration from first-order averaging

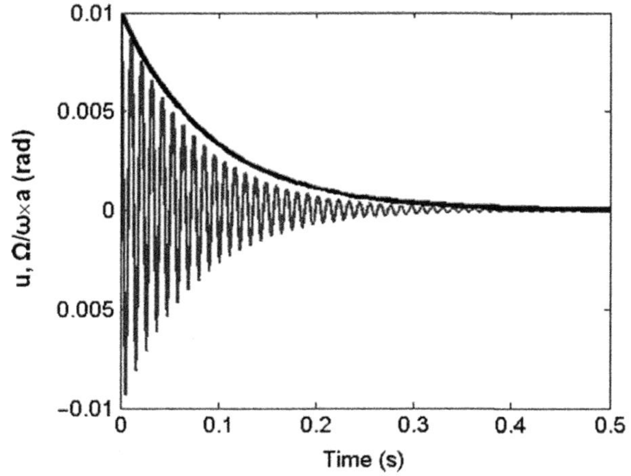

Fig. 6.10 First-order averaging results. $c = 3 \times 10^{-4}$; *grey*: original equation of motion; *black*: amplitude of vibration from first-order averaging

Hopf bifurcation of the original system [69]. The unstable branch, shown by the dotted line, determines the domain of attraction of the trivial or steady-sliding equilibrium point. As discussed by Remark 6.6, for $c > c_{st}$ the steady-sliding state is stable and no limit cycles exist.

Figures 6.12 and 6.13 show the effect of the Stribeck friction (μ_2) and the linear negative friction (μ_3) parameters on the amplitude bifurcation diagram, respectively. In these figures, bifurcation plots are drawn with respect to the applied axial force, R, as the control parameter. As shown, μ_2 controls the domain of attraction of the stable trivial solution without significant change to the limiting value of R. The

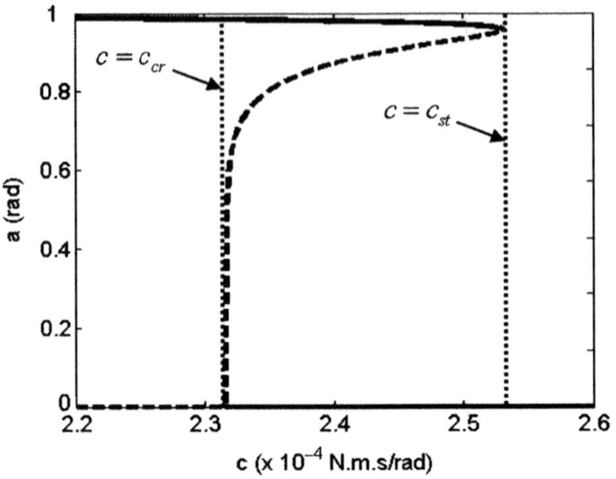

Fig. 6.11 Bifurcation diagram of the averaged amplitude equation. *Solid line* indicates stable; *dashed line* indicates unstable

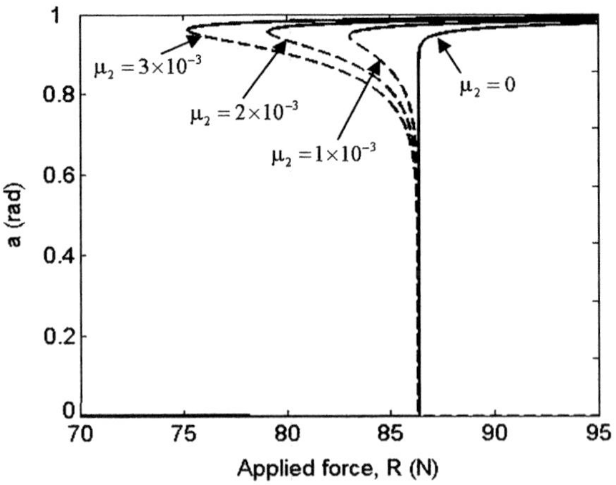

Fig. 6.12 Effect of Stribeck friction on bifurcation

reason for this is that the term $\mu_2 e^{-r_0\Omega}$ is negligible for the considered values of r_0 and Ω [see (6.47)]. However, μ_3 directly controls the threshold of instability of the trivial or steady-state solution.

6.3.7 Numerical Simulation Results: Part 2

In this part, three examples are presented that correspond to the Cases 3 and 4 above. In these examples, the first-order averaged amplitude results are used to study the variations of the amplitude of the steady-state periodic solutions as the input velocity set point (Ω) is varied. In each example, the parameter values not given are those listed in Table 6.2.

Example 1. In this example, $\mu_1 = 0.2$, $\mu_3 = 0$, $r_0 = 0.25$ s/rad, and $r_1 = 2$ s/rad. Three different values are considered for $\mu_2 = 0.01, 0.05, 0.1$. The lead screw support damping is chosen as $c = 1 \times 10^{-4}$ N m s/rad. For each value of μ_2, the coefficient of friction as a function of angular velocity is plotted in Fig. 6.14

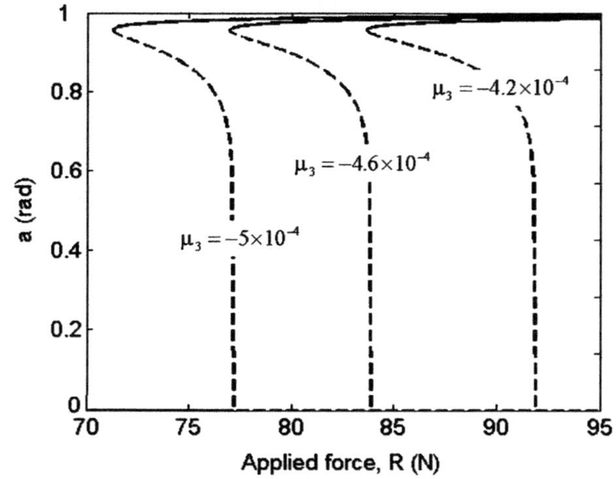

Fig. 6.13 Effect of negative linear friction coefficient, μ_3, on bifurcation

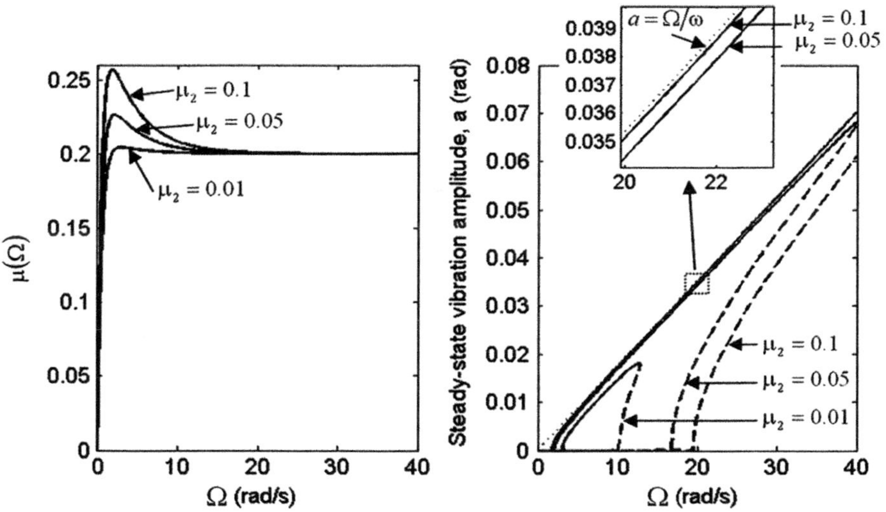

Fig. 6.14 Results for the first example. *Left*: variation of the coefficient of friction with velocity; *right*: variation of steady-state vibration amplitude with input angular velocity

(left). As shown in the steady-state vibration amplitude plots in Fig. 6.14 (right), for the selected value of the damping coefficient, as friction reaches the maximum, the gradient becomes negative ($\Omega > \omega_b$). At this point, the trivial solution loses its stability and a stable limit cycle emerges. For smaller values of μ_2, the region of instability of the trivial solution is smaller.

As shown in the close-up view, larger values of μ_2 result in the stable amplitude of vibrations closer to the limiting value (for the validity of approximations) of $a_{max} = 1$ (or $\alpha_{max} = \Omega/\omega$). For even larger values of μ_2 (not shown), the nontrivial solution of the amplitude equation (corresponding to the stable limit cycle), becomes greater than 1. In these cases, $O(\varepsilon)$ accurate averaging results can be found by using (6.35) or (6.33).

It is interesting to note that, in this example as well as the two examples that follow, as Ω is gradually increased, the trivial equilibrium point first goes through a supercritical pitchfork bifurcation at $\Omega = \omega_b$ and then a subcritical pitchfork bifurcation at $\Omega = \omega^*$ where ω^* is the solution of $b_0 = 0$. These bifurcations in the amplitude equation correspond to Hopf bifurcations of the original system's trivial equilibrium point.

Example 2. Figure 6.15 shows the results for $\mu_1 = 0.2$, $\mu_2 = 0.1$, $\mu_3 = -4 \times 10^{-4}$ s/rad, $r_0 = 0.25$ s/rad, and $r_1 = 2$ s/rad. Three different values are considered here for the lead screw damping; $c = 3 \times 10^{-4}$, 5×10^{-4}, and 7×10^{-4} N m s/rad. As expected, the trivial solution is unstable for $\omega_b < \Omega < \omega^*$ where $\omega_b \approx 1.76$ rad/s and ω^* is the solution of $b_0(\Omega) = 0$. As shown, by increasing the damping, the region of instability of the origin decreases.

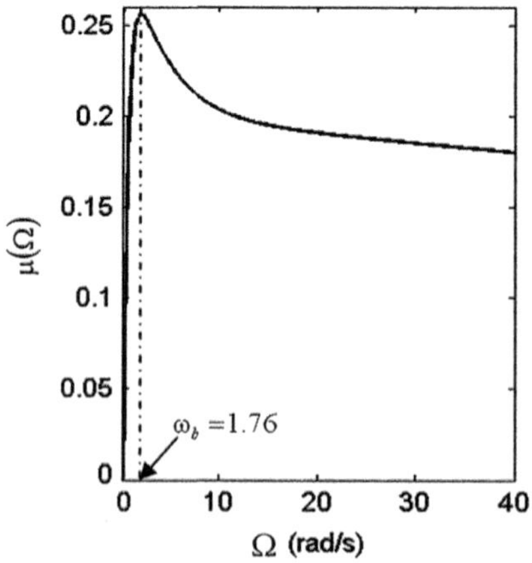

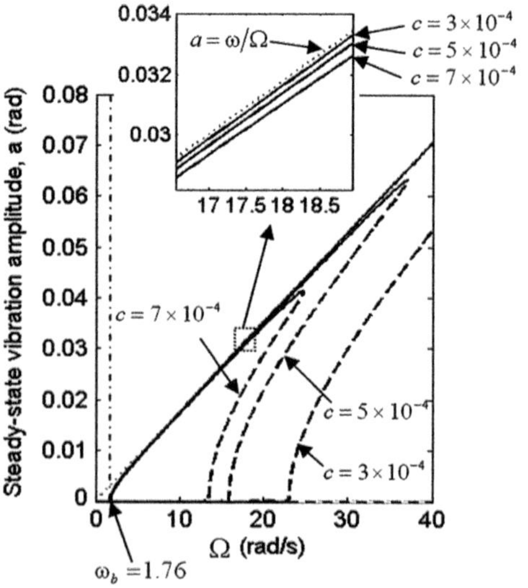

Fig. 6.15 Results for the second example. *Left*: variation of the coefficient of friction with velocity; *right*: variation of steady-state vibration amplitude with input angular velocity

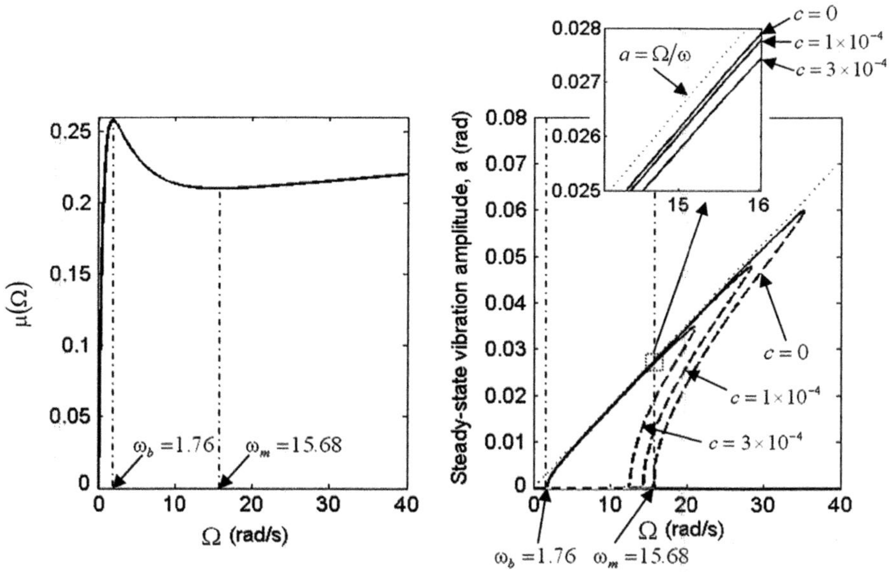

Fig. 6.16 Results for the third example. *Left*: variation of the coefficient of friction with velocity; *right*: variation of steady-state vibration amplitude with input angular velocity

Example 3. Figure 6.16 shows the results for $\mu_1 = 0.2$, $\mu_2 = 0.1$, $\mu_3 = 4 \times 10^{-4}$ s/rad, $r_0 = 0.25$ s/rad, and $r_1 = 2$ s/rad. Three different values are considered here for the lead screw damping; $c = 0$, 1×10^{-4}, and 3×10^{-4} N m s/rad. Once again, the trivial solution is unstable for $\omega_b < \Omega < \omega^*$ where $\omega_b \approx 1.76$ rad/s. For $c = 0$, the trivial solution becomes stable again when $\Omega > \omega_m \approx 15.68$. ω_m corresponds to the local minimum of the friction curve shown in Fig. 6.16 (left).

6.4 Conclusions

In this chapter, the 1-DOF model of the lead screw drives developed in Sect. 5.3 was used to study the instability caused by the negative gradient of the friction coefficient with respect to velocity. The local stability of the steady-sliding equilibrium point of the system was studied by examining the eigenvalues of the Jacobian matrix of the linearized system. It was shown that the steady-sliding equilibrium point of the system loses stability when the condition given by (6.9) is satisfied.

The eigenvalue analysis result was extended by the application of the method of averaging. It was shown that depending on the system parameters, one of the following cases define the dynamic behavior of the system:

1. The trivial solution is stable and no other solution exists.
2. The trivial solution is stable and it is surrounded by an unstable limit cycle that defines the region of attraction of the trivial solution. The unstable limit cycle is

inside a stable limit cycle. The presence of Stribeck effect is a necessary condition in this scenario.

3. The trivial solution is unstable and it is surrounded by a stable limit cycle.

The numerical simulation results presented, also showed the applicability of the averaging results in approximating the amplitude of periodic vibrations. The accuracy of the first-order approximations can be improved by using higher-order averaging. In Appendix B, equations for the second- and third-order averaging is derived for the 1-DOF lead screw model studied in this chapter. The improved accuracy of the approximate periodic solution is shown by a numerical example. In Appendix C, the first-order averaging method is applied to the 2-DOF model of Sect. 5.6 to study the negative damping instability mechanism.

Chapter 7
Mode Coupling Instability Mechanism

In this chapter, the mode coupling instability in the lead screw drives is studied. As mentioned in Sect. 4.2, mode coupling is exclusive to multi-DOF systems and can destabilize a system even when the coefficient of friction is independent of sliding velocity.

The two 2-DOF models of Sects. 5.5 and 5.6 as well as the 3-DOF model of Sect. 5.7 are studied in this chapter. First, in Sect. 7.1, the linearized models of the 2-DOF models of the lead screw are derived, and the role of friction in breaking the symmetry of the system inertia and stiffness matrices is demonstrated. In Sect. 7.2, the eigenvalue analysis method is used to study the stability of the linearized 2-DOF system with compliant threads under the action of the mode coupling instability mechanism. In this section, undamped and damped cases are treated separately. In Sect. 7.3, similarities between the two system models presented in Sect. 7.1 in terms of the stability conditions of the steady-sliding equilibrium point are studied. To gain some insight into the nonlinear behavior of the lead screw system under mode coupling instability, a series of numerical simulation results are presented and discussed in Sect. 7.4. The mode coupling instability in a 3-DOF lead screw model is briefly discussed in Sect. 7.5. The conclusions drawn in this chapter are summarized in Sect. 7.6.

7.1 Mathematical Models

In Sects. 5.5 and 5.6, we have introduced two 2-DOF models for the lead screw drives. In this section, the equations of motion of these models are transformed into matrix form and linearized with respect to their respective steady-sliding equilibrium point. These equations are then used in the next sections to study the local stability of the equilibrium point and the role of mode coupling instability mechanism. In this chapter, for simplicity, the coefficient of friction, μ, is taken as a constant.

O. Vahid-Araghi and F. Golnaraghi, *Friction-Induced Vibration in Lead Screw Drives*,
DOI 10.1007/978-1-4419-1752-2_7, © Springer Science+Business Media, LLC 2011

7.1.1 2-DOF Model with Axially Compliant Lead Screw Supports

The equations of motion of the 2-DOF lead screw drive with axially compliant lead screw support are given in Sect. 5.6 by (5.30) and (5.31). Assume $\theta_i = \Omega t$ where Ω is a constant. The following change of variables is used to transfer the steady-sliding equilibrium point to the origin:

$$\begin{aligned} \theta &= y_1 + \theta_i + u_{10}, \\ x_1 &= r_m \tan\lambda y_2 + u_{20}, \end{aligned} \tag{7.1}$$

where

$$\begin{aligned} u_{10} &= \frac{-c\Omega - r_m\xi_0[R - F_0\mathrm{sgn}(\Omega)] - T_0\mathrm{sgn}(\Omega)}{k}, \\ u_{20} &= \frac{R - F_0\mathrm{sgn}(\Omega)}{k_1}, \end{aligned} \tag{7.2}$$

where

$$\xi_0 = \frac{\mu\mathrm{sgn}(R\Omega) - \tan\lambda}{1 + \mu\mathrm{sgn}(R\Omega)\tan\lambda}. \tag{7.3}$$

Substituting (7.1) into (5.30) and (5.31) and rearranging yield

$$\tilde{\mathbf{M}}_1\ddot{\mathbf{y}} + \mathbf{C}_1\dot{\mathbf{y}} + \mathbf{K}_1\mathbf{y} = \mathbf{f}_1(\mathbf{y}, \dot{\mathbf{y}}), \tag{7.4}$$

where

$$\tilde{\mathbf{M}}_1 = \begin{bmatrix} I - \xi\tan\lambda m{r_m}^2 & -\xi\tan\lambda m{r_m}^2 \\ \tan^2\lambda m{r_m}^2 & \tan^2\lambda(m + m_1){r_m}^2 \end{bmatrix}, \tag{7.5}$$

$$\mathbf{C}_1 = \begin{bmatrix} c & 0 \\ 0 & c_1{r_m}^2\tan^2\lambda \end{bmatrix},$$

$$\mathbf{K}_1 = \begin{bmatrix} k & 0 \\ 0 & k_1{r_m}^2\tan^2\lambda \end{bmatrix}, \tag{7.6}$$

$$\mathbf{f}_1 = \begin{bmatrix} r_m(\xi_0 - \xi)R - r_mF_0(\xi_0\mathrm{sgn}(\Omega) - \xi\mathrm{sgn}(\dot{\theta})) + T_0(\mathrm{sgn}(\Omega) - \mathrm{sgn}(\dot{\theta})) \\ r_mF_0\tan\lambda(\mathrm{sgn}(\Omega) - \mathrm{sgn}(\dot{\theta})) \end{bmatrix}. \tag{7.7}$$

Note that the nonlinear force vector given by (7.7) is nonzero only when a trajectory reaches (or crosses) the stick-slip boundary (i.e., $\text{sgn}(\dot{y}_1 + \Omega) \neq \text{sgn}(\Omega)$) or when the contact force changes sign (i.e., $\text{sgn}(N) \neq \text{sgn}(R)$). The linearized system equation in matrix form is given by

$$\mathbf{M}_1\ddot{\mathbf{y}} + \mathbf{C}_1\dot{\mathbf{y}} + \mathbf{K}_1\mathbf{y} = 0, \tag{7.8}$$

where

$$\mathbf{M}_1 = \begin{bmatrix} I - \xi_0 \tan\lambda m r_\mathrm{m}{}^2 & -\xi_0 \tan\lambda m r_\mathrm{m}{}^2 \\ \tan^2\lambda m r_\mathrm{m}{}^2 & \tan^2\lambda (m + m_1) r_\mathrm{m}{}^2 \end{bmatrix}. \tag{7.9}$$

Remark 7.1. In the absence of friction, the inertia matrix (7.5) simplifies to

$$\mathbf{M}_1\big|_{\mu=0} = \begin{bmatrix} I + \tan^2\lambda m r_\mathrm{m}{}^2 & \tan^2\lambda m r_\mathrm{m}{}^2 \\ \tan^2\lambda m r_\mathrm{m}{}^2 & \tan^2\lambda (m + m_1) r_\mathrm{m}{}^2 \end{bmatrix},$$

which is, of course, symmetric. However, when $\mu \neq 0$, $\mathbf{M}_1$ is asymmetric. □

7.1.2 2-DOF Model with Compliant Threads

The equations of motion of the 2-DOF lead screw drive with compliant threads are given in Sect. 5.5 by (5.25) and (5.26). Once again, assume $\theta_\mathrm{i} = \Omega t$ where Ω is a constant. The following change of variables is used to transfer the steady-sliding equilibrium point to the origin:

$$\begin{aligned} \theta &= y_1 + \theta_\mathrm{i} + u_{10}, \\ x &= r_\mathrm{m} \tan\lambda y_2 + u_{20} + r_\mathrm{m} \tan\lambda \theta_\mathrm{i}, \end{aligned} \tag{7.10}$$

where u_{10} is given by (7.2) and

$$u_{20} = \frac{1}{k_\mathrm{c}\cos^2\lambda} \frac{R - F_0 \,\text{sgn}(\Omega)}{1 + \mu \,\text{sgn}(R\Omega) \tan\lambda} + r_\mathrm{m} \tan\lambda u_{10}.$$

Substituting (7.10) into (5.25) and (5.26) and rearranging yield

$$\mathbf{M}_2\ddot{\mathbf{y}} + \mathbf{C}_2\dot{\mathbf{y}} + \mathbf{K}_2\mathbf{y} = \mathbf{f}_2(\mathbf{y}, \dot{\mathbf{y}}), \tag{7.11}$$

where

$$\mathbf{M}_2 = \begin{bmatrix} I & 0 \\ 0 & \hat{m} \end{bmatrix},$$

$$\mathbf{K}_2 = \begin{bmatrix} k + \hat{k}_c - \mu \mathrm{sgn}(R\Omega) \cot \lambda \hat{k}_c & -\hat{k}_c + \mu \mathrm{sgn}(R\Omega) \cot \lambda \hat{k}_c \\ -\hat{k}_c - \mu \mathrm{sgn}(R\Omega) \tan \lambda \hat{k}_c & \hat{k}_c + \mu \mathrm{sgn}(R\Omega) \tan \lambda \hat{k}_c \end{bmatrix}, \tag{7.12}$$

$$\mathbf{C}_2 = \begin{bmatrix} c + \hat{c}_c - \mu \mathrm{sgn}(R\Omega) \cot \lambda \hat{c}_c & -\hat{c}_c + \mu \mathrm{sgn}(R\Omega) \cot \lambda \hat{c}_c \\ -\hat{c}_c - \mu \mathrm{sgn}(R\Omega) \tan \lambda \hat{c}_c & \hat{c}_c + \mu \mathrm{sgn}(R\Omega) \tan \lambda \hat{c}_c \end{bmatrix}, \tag{7.13}$$

where

$$\hat{k}_c = {r_m}^2 \sin^2 \lambda k_c, \tag{7.14}$$

$$\hat{c}_c = {r_m}^2 \sin^2 \lambda c_c, \tag{7.15}$$

$$\hat{m} = {r_m}^2 \tan^2 \lambda m. \tag{7.16}$$

The force vector, $\mathbf{f}_2$, is found as

$$\mathbf{f}_2(\mathbf{y}, \dot{\mathbf{y}}) = \mu(\mathrm{sgn}(R\Omega) - \mathrm{sgn}(N(\dot{y}_1 + \Omega))) \left(\hat{k}_c y_2 - \hat{k}_c y_1 + \hat{c}_c \dot{y}_2 - \hat{c}_c \dot{y}_1 + \hat{R}\right) \begin{bmatrix} \cot \lambda \\ \tan \lambda \end{bmatrix} + T_0 \begin{bmatrix} \mathrm{sgn}(\Omega) - \mathrm{sgn}(\dot{y}_1 + \Omega) \\ 0 \end{bmatrix}, \tag{7.17}$$

where

$$\hat{R} = \frac{r_m \tan \lambda}{1 + \mu \mathrm{sgn}(R\omega) \tan \lambda} [R - F_0 \mathrm{sgn}(\Omega)].$$

Similar to (7.7), the nonlinear force vector given by (7.17) is nonzero only when $\mathrm{sgn}(\dot{y}_1 + \Omega) \neq \mathrm{sgn}(\Omega)$ or $\mathrm{sgn}(N) \neq \mathrm{sgn}(R)$. The linearized system equation in matrix form is given by

$$\mathbf{M}_2 \ddot{\mathbf{y}} + \mathbf{C}_2 \dot{\mathbf{y}} + \mathbf{K}_2 \mathbf{y} = 0. \tag{7.18}$$

Remark 7.2. In the absence of friction, the stiffness (7.12) and damping (7.13) matrices simplify to

$$\mathbf{K}_2|_{\mu=0} = \begin{bmatrix} k + \hat{k}_c & -\hat{k}_c \\ -\hat{k}_c & \hat{k}_c \end{bmatrix} \quad \text{and} \quad \mathbf{C}_2 = \begin{bmatrix} c + \hat{c}_c & -\hat{c}_c \\ -\hat{c}_c & \hat{c}_c \end{bmatrix},$$

which are symmetric. However, when $\mu \neq 0$, $\mathbf{K}_2$ and $\mathbf{C}_2$ are asymmetric. □

7.2 Linear Stability Analysis

The two linear systems given by (7.8) and (7.18) share one very important feature: not all coefficient matrices are symmetric. The asymmetry, which is caused by friction, may lead to flutter instability (also known as mode coupling). The system defined by (7.8) may also lose stability due to kinematic constraint instability mechanism.[1]

We start by investigating the possibility of divergence instability. From the discussions of Chap. 2, we know that at the divergence instability boundary, the stiffness matrix becomes singular (i.e., $\det(\mathbf{K}_i) = 0, \quad i = 1, 2$). From (7.6) and (7.12), one may find

$$\det(\mathbf{K}_1) = kk_1{r_m}^2\tan^2\lambda,$$
$$\det(\mathbf{K}_2) = k\hat{k}_c(1 + \mu\,\mathrm{sgn}(R\Omega)\tan\lambda),$$

which are always positive.[2] Hence, divergence is ruled out for the steady-sliding equilibrium points of the two 2-DOF lead screw models presented above.

The remainder of this chapter is dedicated to the mode coupling instability mechanism and is focused on the linear system given by (7.18). In Sect. 7.3, the similarities between the two models are explored. The undamped case will be treated first and then the effect of damping is studied.

7.2.1 Undamped System

Setting $\mathbf{C}_2 = 0$ in (7.18), the linearized model of the undamped lead screw drive model with compliant threads is obtained as

$$\mathbf{M}_2\ddot{\mathbf{y}} + \mathbf{K}_2\mathbf{y} = 0. \tag{7.19}$$

The natural frequencies of this system are the roots of the following equation:

$$\det\left(\mathbf{K}_2 - \omega^2\mathbf{M}_2\right) = 0, \tag{7.20}$$

which is a quadratic equation in ω^2. Expanding (7.20) yields

$$a_4\omega^4 + a_2\omega^2 + a_0 = 0,$$

[1]See Sect. 8.8.1 for the analysis of kinematic constraint instability in this system.

[2]It is assumed that the condition; $\mu < \cot\lambda$ always holds. Violation of this condition would require a very high coefficient of friction or a lead screw with a helix angle greater than 45°, which are not encountered in practical situation [33].

where

$$a_4 = \hat{m}I,$$
$$a_2 = \hat{k}_c\hat{m}\cot\lambda(\mu\mathrm{sgn}(R\Omega) - \tan\lambda) - \hat{k}_c I(1 + \mu\mathrm{sgn}(R\Omega)\tan\lambda) - k\hat{m},$$
$$a_0 = k\hat{k}_c(1 + \mu\mathrm{sgn}(R\Omega)\tan\lambda).$$

Since $a_4 > 0$ and $a_0 > 0$, instability occurs whenever

$$a_2 > 0 \tag{7.21}$$

or

$$a_2{}^2 - 4a_0a_4 < 0 \tag{7.22}$$

Instability condition given by (7.21) can be rearranged as

$$a_2 = -(1 + \mu\mathrm{sgn}(R\Omega)\tan\lambda)\hat{k}_c\Gamma_0 - k\hat{m} > 0, \tag{7.23}$$

where

$$\Gamma_0 = I - \xi_0 \tan\lambda m r_m{}^2 > 0$$

and (7.16) was used. Also, ξ_0 is given by (7.3). Obviously, if $\Gamma_0 > 0$, inequality (7.23) cannot be satisfied. On the other hand, if $\mathrm{sgn}(R\omega) = 1$ and $\mu > \tan\lambda$, then for suitable values of system parameters, the inequality $\Gamma_0 < 0$ is satisfied. In this case, the origin is unstable if

$$-\cos^2\lambda(1 + \mu\tan\lambda)\frac{\Gamma_0}{m} > \frac{k}{k_c}, \tag{7.24}$$

where (7.14) and (7.16) were used. However, since $a_2 = 0$ satisfies (7.22), the boundary that is defined by $a_2 = 0$ is inside the region defined by (7.22) in the system parameters space. Consequently, $a_2 = 0$ does not define the boundary for the initial loss of stability.

The second instability condition, given by (7.22), represents the mode coupling (flutter) instability. The equation for the flutter instability boundary (i.e., coalescence of the two real natural frequencies) is found by replacing the less than sign with the equal sign in (7.22). After some manipulations, one finds

$$b_1\hat{k}_c{}^2 + b_2k\hat{k}_c + b_3k^2 = 0, \tag{7.25}$$

where

$$\begin{aligned} b_1 &= [\hat{m}(1 - \mu\mathrm{sgn}(R\Omega)\cot\lambda) + I(1 + \mu\mathrm{sgn}(R\Omega)\tan\lambda)]^2, \\ b_2 &= 2\hat{m}[\hat{m}(1 - \mu\mathrm{sgn}(R\Omega)\cot\lambda) - I(1 + \mu\mathrm{sgn}(R\Omega)\tan\lambda)], \\ b_3 &= \hat{m}^2. \end{aligned} \tag{7.26}$$

This equation is quadratic in k and $\hat{k}_c$ and can be solved to find parametric relationships for the onset of the flutter instability (either for k as a function of $\hat{k}_c$ or vice versa). The conditions for the existence of positive real solutions to (7.25) are given by the following lemma.

Lemma 7.1. *The necessary conditions for the mode coupling instability in the model defined by (7.19) are given as follows:*

$$\mu > \tan\lambda \quad \wedge \quad R\Omega > 0. \tag{7.27}$$

Proof. The conditions for the solutions of (7.25) (for either k or $\hat{k}_c$) to be real positive numbers are

$$b_2{}^2 - 4b_1b_3 \geq 0, \tag{7.28}$$

$$b_2 < 0. \tag{7.29}$$

Using (7.26), the inequality (7.28) can be expressed as

$$16\hat{m}^3 I(1 - \mu\mathrm{sgn}(R\Omega)\cot\lambda)(1 + \mu\mathrm{sgn}(R\Omega)\tan\lambda) \leq 0,$$

which holds if and only if $\xi_0 > 0$. This in turn requires (7.27). The second inequality, (7.29), can be expressed as

$$(1 + \mu\mathrm{sgn}(R\Omega)\tan\lambda)\left(I + \tan\lambda\xi_0 m r_\mathrm{m}{}^2\right) > 0,$$

which is satisfied for $\xi_0 > 0$. ■

Remark 7.3. Lemma 7.1 establishes that for the undamped system (7.19), mode coupling instability can only occur when the lead screw drive is self-locking and the applied force is in the direction of the translation. □

7.2.2 Examples and Discussion

The parameter values used in the subsequent numerical examples are listed in Table 7.1. First, notice that the self-locking condition is satisfied for the selected value of the constant coefficient of friction (i.e., $\mu = 0.218 > \tan(5.57°) = 0.0975$). Also $R\Omega > 0$.

Figure 7.1 shows the stability region of the undamped 2-DOF model in the $k_c - \mu$ parameter space. The hatched region corresponds to the parameter range where the two natural frequencies are complex, and the steady-sliding equilibrium point is unstable. The boundary of this region is the flutter instability threshold where $\omega_1 = \omega_2$.

Table 7.1 Parameter values used in the examples

Parameter	Value	Parameter	Value
r_m	5.18 mm	c	20×10^{-5} N m s/rad
λ	5.57°	R	100 N
I	3.12×10^{-6} kg m^2	μ	0.218
k	1 N m/rad	Ω	40 rad/s
T_0	0 N m	F_0	0 N

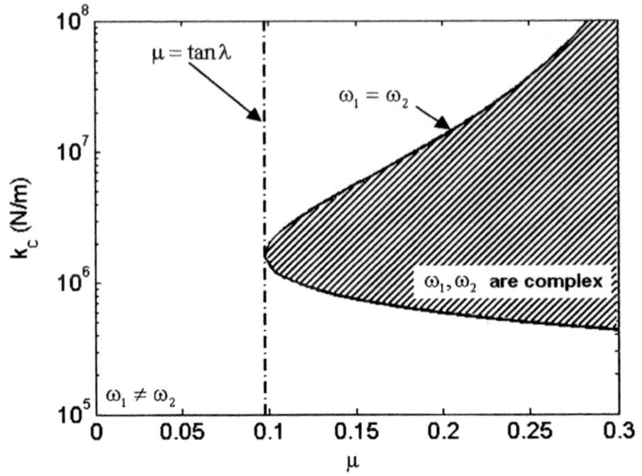

Fig. 7.1 Stability of the 2-DOF system as support stiffness and coefficient of friction are varied. $m = 5$. *Hatched region*: unstable

It is interesting to note that the flutter boundary is tangent to the $\mu = \tan\lambda$ line (at $k_c \approx 1.7 \times 10^6$). As predicted, the instability region lies entirely to the right of this line. Figure 7.2 shows the variation of the real and imaginary parts of the eigenvalues of the undamped system for $\mu = 0.2 > \tan\lambda$. It can be seen that flutter instability occurs as the two natural frequencies merge. Further increase of the contact stiffness uncouples the two modes and the stability is restored.

For larger values of the translating mass, m, inequality (7.21) can also become active in the considered ranges of the parameters. An example of this situation is given by the stability map of Fig. 7.3. The translating mass is increased to $m = 15$ kg, resulting in Γ_0 vanishing at $\mu \approx 0.178$. The hatched regions in this figure show parameter ranges where these two mechanisms are active.

Consider the variation of the system parameters μ and k_c along the dashed line in Fig. 7.3. In the stable region, the two frequencies are distinct real numbers. At the flutter instability boundary, the two frequencies merge, i.e., ${\omega_{1,2}}^2 = a, \quad a > 0$. By further increasing the parameters, the frequencies become complex valued, i.e., ${\omega_{1,2}}^2 = a \pm ib, \quad a, b > 0$. At the boundary $a_2 = 0$, the real part of ${\omega_{1,2}}^2$ vanishes, i.e., ${\omega_{1,2}}^2 = \pm\, ib, \quad b > 0$. If the parameters are increased even further, the real part of the squared frequencies becomes negative, i.e., ${\omega_{1,2}}^2 = -a \pm ib, \quad a, b > 0$.

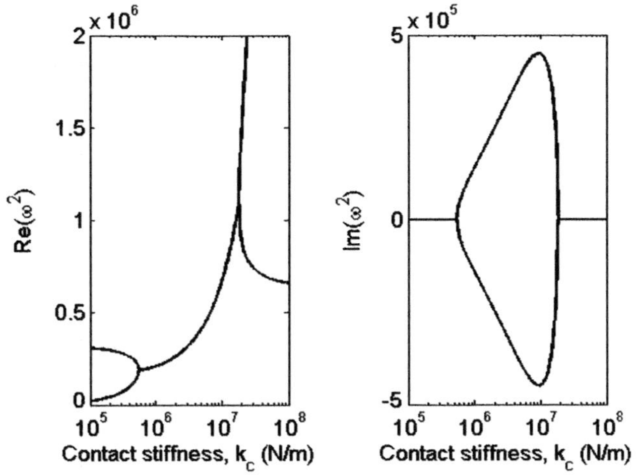

Fig. 7.2 Variation of the real and imaginary parts of the eigenvalues as the contact stiffness is varied. $\mu = 0.2$, $m = 5$

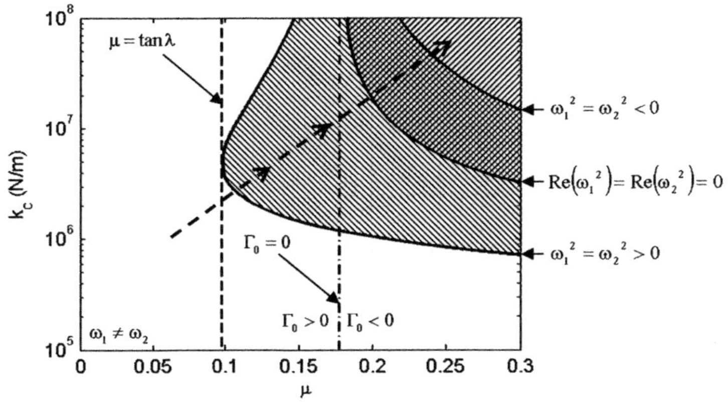

Fig. 7.3 Stability of the 2-DOF system as contact stiffness and coefficient of friction are varied, when $m = 15$ and $R\omega > 0$. The hatched area: mode coupling instability region; the hatched area: (7.21) is satisfied

At the second boundary of flutter instability, the squared frequencies are identical and purely imaginary, i.e., ${\omega_{1,2}}^2 = -a, \quad a > 0$. Beyond this threshold, squared frequencies are different but remain purely imaginary.

The variation of the real and imaginary parts of eigenvalues (i.e., natural frequencies) for $\mu = 0.218 > 0.178$ as k_c is varied is plotted in Fig. 7.4. Mode coupling instability occurs at the flutter boundary: $k_c = 9.65 \times 10^5$. It is interesting to note that, due to the activation of (7.21) inequality, the origin remains unstable even when k_c is large enough that the two modes decouple.

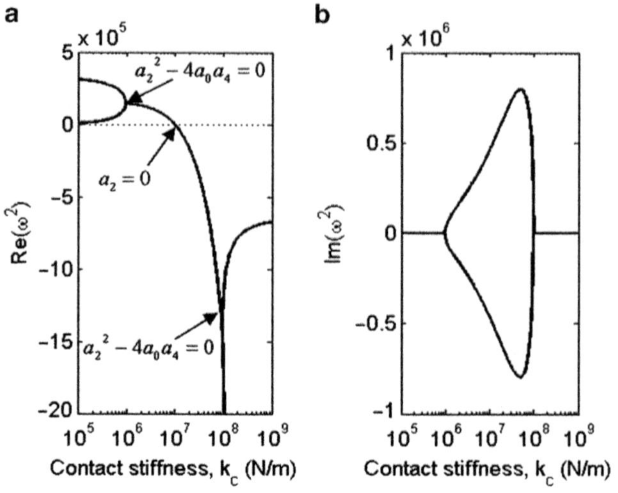

Fig. 7.4 Variation of the real part (**a**) and imaginary part (**b**) of the eigenvalues as the contact stiffness is varied. $\mu = 0.218$, $m = 15$

7.2.3 Damped System

The eigenvalues of the damped 2-DOF system ($\eta_i, i = 1\ldots4$) are the solutions of the fourth-order equation

$$\det\left(\eta^2\mathbf{M}_2 + \eta\mathbf{C}_2 + \mathbf{K}_2\right) = 0.$$

Assuming all of the system parameters to be non-negative numbers, the stability conditions based on the Routh–Hurwitz criterion are found to be

$$D_1 = I\hat{m} > 0, \tag{7.30}$$

$$D_2 = \hat{c}_c\Delta + c\hat{m} > 0, \tag{7.31}$$

$$D_3 = \hat{k}_c\Delta + k\hat{m} + c\hat{c}_c(1 + \mu\mathrm{sgn}(R\Omega)\tan\lambda) > 0, \tag{7.32}$$

$$D_4 = D_2D_3 - \hat{m}I\left(c\hat{k}_c + \hat{c}_ck\right)(1 + \mu\mathrm{sgn}(R\Omega)\tan\lambda) > 0, \tag{7.33}$$

$$D_4\left(c\hat{k}_c + \hat{c}_ck\right) - k\hat{k}_cD_2{}^2 > 0, \tag{7.34}$$

where

$$\Delta = I(1 + \mu\mathrm{sgn}(R\Omega)\tan\lambda) + \hat{m}(1 - \mu\mathrm{sgn}(R\Omega)\cot\lambda). \tag{7.35}$$

Obviously, the condition (7.30) is always satisfied. To find the instability boundaries, the greater than sign in inequalities (7.31)–(7.34) are changed to equal sign.

Lemma 7.2. *For the linear system described by (7.18), the initial instability boundary is defined by*

$$D_4\left(c\hat{k}_c + \hat{c}_c k\right) - k\hat{k}_c D_2^{\,2} = 0, \tag{7.36}$$

where D_2 and D_4 are given by (7.31) and (7.33), respectively.

Proof. If $D_4 = 0$, (7.34) is violated. If $D_2 = 0$ or $D_3 = 0$, (7.33) is violated. Thus, none of these equations define the initial instability boundary which leaves only (7.36). ∎

The equation (7.36) is too complicated to be useful in establishing closed-form parametric stability boundaries. However, some important special cases can be proven from the above Routh–Hurwitz conditions which are listed here in the form of simple lemmas.

Lemma 7.3. *The equilibrium point of the linear system described by (7.18) is stable when the force is applied opposite to the nut translation direction, i.e., $R\Omega < 0$.*

Proof. To show that the Routh-Hurwitz conditions hold, we only need to show that (7.31), (7.32), and (7.34) hold. Note that (7.30) is always true and (7.33) follows from (7.34). Since $R\Omega > 0$ we have $\Delta = I(1 - \mu\tan\lambda) + \hat{m}(1 + \mu\cot\lambda) > 0$ thus $D_2 > 0$ and $D_3 > 0$. Substituting (7.31)–(7.33) and (7.35) into the expression of (7.34) and using $\operatorname{sgn}(R\Omega) = -1$ give (after some algebra),

$$\begin{aligned}
D_4\left(c\hat{k}_c + \hat{c}_c k\right) - k\hat{k}_c D_2^{\,2} = {} & c\hat{c}_c\left(\hat{k}_c I(1-\mu\tan\lambda) - k\hat{m}\right)^2 \\
& + c\hat{c}_c\hat{k}_c^{\,2}\left(\hat{m}^2(1+\mu\cot\lambda)^2 + 2\hat{m}I(1-\mu\tan\lambda)(1+\mu\cot\lambda)\right) \\
& + \left(\hat{c}_c^{\,2}k^2 + c^2\hat{k}_c^{\,2}\right)\hat{m}^2(1+\mu\cot\lambda) \\
& + c\hat{c}_c(\hat{c}_c\Delta + c\hat{m})\left(c\hat{k}_c + \hat{c}_c k\right)(1-\mu\tan\lambda) > 0,
\end{aligned}$$

which is positive for $\mu < \cot\lambda$. ∎

Lemma 7.4. *The equilibrium point of the linear system described by (7.18) is stable when the force is applied in the direction of the nut translation and the lead screw drive is not self-locking, i.e., $R\Omega > 0$ and $\mu < \tan\lambda$.*

Proof. Since $R\Omega < 0$ and $\mu < \tan\lambda$ we have $\Delta = I(1 + \mu\tan\lambda) + \hat{m}(1 - \mu\cot\lambda) > 0$ thus $D_2 > 0$ and $D_3 > 0$. Similar to the previous lemma after substituting (7.31)–(7.33) and (7.35) into the expression of (7.34) gives

$$\begin{aligned}
D_4\left(c\hat{k}_c + \hat{c}_c k\right) - k\hat{k}_c D_2^{\,2} = {} & c\hat{c}_c\left(\hat{k}_c I(1+\mu\tan\lambda) - k\hat{m}\right)^2 \\
& + c\hat{c}_c\hat{k}_c^{\,2}\left(\hat{m}^2(1-\mu\cot\lambda)^2 + 2\hat{m}I(1+\mu\tan\lambda)(1-\mu\cot\lambda)\right) \\
& + \left(\hat{c}_c^{\,2}k^2 + c^2\hat{k}_c^{\,2}\right)\hat{m}^2(1-\mu\cot\lambda) \\
& + c\hat{c}_c(\hat{c}_c\Delta + c\hat{m})\left(c\hat{k}_c + \hat{c}_c k\right)(1+\mu\tan\lambda),
\end{aligned} \tag{7.37}$$

which is positive for $\mu < \tan\lambda$. ∎

Remark 7.4. Similar to the undamped case (Lemma 7.1), the self-locking and the application of the axial load in the direction of travel are the two necessary (but not sufficient) conditions for the instability to occur. □

Lemma 7.5. *The equilibrium point of the linear system described by (7.18) is unstable when* $R\Omega > 0$, $\mu > \tan\lambda$, *and* $c_c = 0 \wedge c \neq 0$.

Proof. setting $c_c = 0$ and $\mathrm{sgn}(R\Omega) = 1$ in (7.37) gives

$$D_4\left(c\hat{k}_c + \hat{c}_c k\right) - k\hat{k}_c D_2{}^2 = c\hat{k}_c D_4 - k\hat{k}_c D_2{}^2 = c^2\hat{k}_c{}^2\hat{m}^2(1 - \mu\cot\lambda),$$

which is negative for $\mu > \tan\lambda$. ■

Lemma 7.6. *The equilibrium point of the linear system described by (7.18) is unstable when* $R\Omega > 0$, $\mu > \tan\lambda$, *and* $c_c \neq 0 \wedge c = 0$.

Proof. setting $c = 0$ and $\mathrm{sgn}(R\Omega) = 1$ in (7.37) gives

$$D_4\left(c\hat{k}_c + \hat{c}_c k\right) - k\hat{k}_c D_2{}^2 = \hat{c}_c k D_4 - k\hat{k}_c D_2{}^2 = \hat{c}_c{}^2 k^2\hat{m}^2(1 - \mu\cot\lambda),$$

which is negative for $\mu > \tan\lambda$. ■

The presence of damping only in the rotational DOF (i.e., lead screw support damping, c) or in the translating DOF (i.e., contact damping, c_c) destabilizes the system. Similar qualitative observations are reported in the literature regarding simple systems (see, e.g., [78, 115–117]).

7.2.4 Examples and Discussion

In Fig. 7.5, the two damping coefficients are chosen as $c_c = 10^2$ and $c = 2 \times 10^{-5}$. Other system parameters are selected as before and $m = 5$. It can be seen that the addition of damping, contrary to common experiences has reduced the stability region.

Variation of the eigenvalues for this case is plotted in Fig. 7.6 for $\mu = 0.15$. The coalescence of the natural frequencies can be seen in this figure. It must be noted that, due to the presence of damping, the two frequencies do not match exactly,[3] and the instability region does not necessarily correspond to the range where they are close.

By increasing the damping, as shown in Fig. 7.7, the stable region is expanded beyond the instability region of the undamped system. In this figure, the damping coefficients are $c_c = 2 \times 10^3$ and $c = 4 \times 10^{-4}$.

[3]See footnote on page 50.

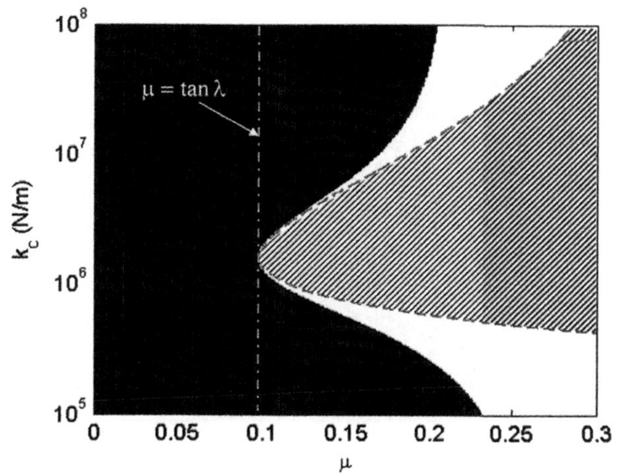

Fig. 7.5 Regions of stability of the 2-DOF model with damping. *Black*: stable, *white*: unstable. $R\Omega > 0$, $m = 5$, $c_c = 10^2$, and $c = 4 \times 10^{-5}$

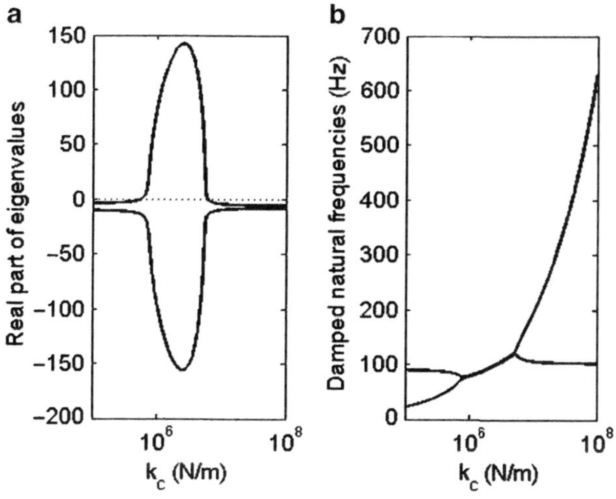

Fig. 7.6 Variation of the real part of the eigenvalues (**a**) and the natural frequencies (**b**), as the contact stiffness is varied. $\mu = 0.15$, $m = 5$, $c_c = 10^2$, and $c = 4 \times 10^{-5}$

Similar to Fig. 7.6, Fig. 7.8 shows the evolution of the real and imaginary parts of the eigenvalues as k_c is varied, for $\mu = 0.15$. The increased damping has resulted in the "overdamping" of the lower mode for roughly $k_c < 1.96 \times 10^5$. Moreover, in this higher damping level, the range over which the two natural frequencies are close has been almost completely eliminated.

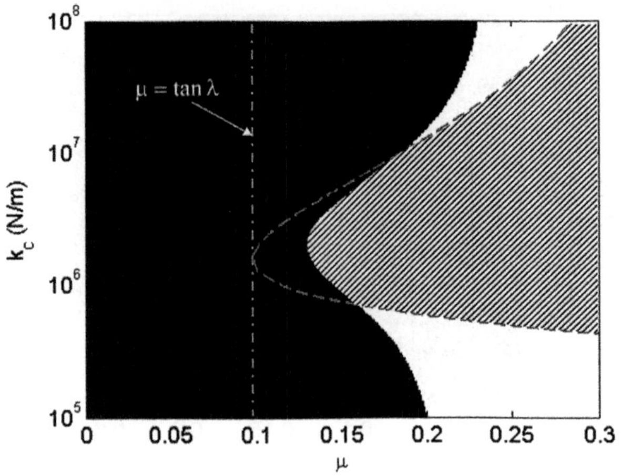

Fig. 7.7 Regions of stability of the 2-DOF model with damping. *Black*: stable, *white*: unstable. $R\Omega > 0$, $m = 5$, $c_c = 2 \times 10^3$, and $c = 4 \times 10^{-4}$

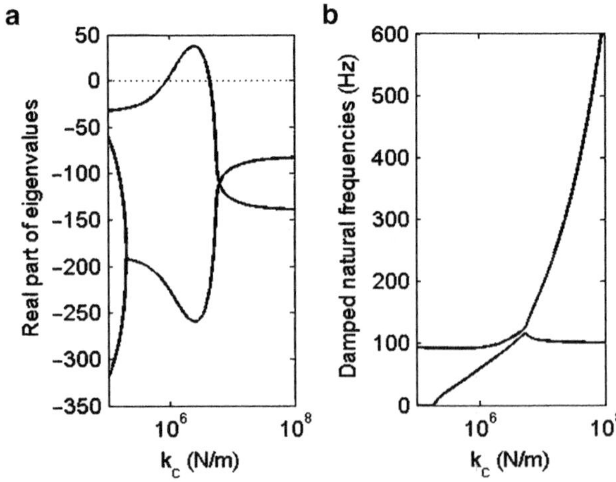

Fig. 7.8 Variation of the real part of the eigenvalues (**a**) and the natural frequencies (**b**), as the contact stiffness is varied. $\mu = 0.15$, $m = 5$, $c_c = 2 \times 10^3$, and $c = 4 \times 10^{-4}$

7.3 Comparison Between the Stability Conditions of the Two Lead Screw Models

In Sect. 7.1, we have seen the role of friction in the two lead screw models through breaking the symmetry of the linearized system inertia, damping, and stiffness matrices. The damping and stiffness matrices of model (7.8) (i.e., 2-DOF lead screw drive model with axially compliant lead screw supports) are symmetric while the inertia matrix is asymmetric. On the other hand, the damping and stiffness matrices of model (7.18) (i.e., 2-DOF lead screw drive model with compliant threads) are asymmetric while the inertia matrix is symmetric. Aside from the possibility of kinematic constraint instability in system (7.8), the linear stability conditions of these two different systems are very similar as it will be demonstrated in the following paragraphs.

The characteristic equation for the linear system (7.8) is found as

$$\det\left(\eta^2 \mathbf{M}_1 + \eta \mathbf{C}_1 + \mathbf{K}_1\right) = 0.$$

Based on this fourth-order equation, the Routh–Hurwitz stability conditions are:

$$D_1 = I(m + m_1) - m m_1 \xi_0 {r_m}^2 \tan\lambda > 0, \tag{7.38}$$

$$D_2 = c(m + m_1) + c_1\left(I - \xi_0 \tan\lambda m {r_m}^2\right) > 0, \tag{7.39}$$

$$D_3 = k(m + m_1) + c c_1 + k_1\left(I - \xi_0 \tan\lambda m {r_m}^2\right) > 0, \tag{7.40}$$

$$D_4 = D_2 D_3 - (k c_1 + c k_1) D_1 > 0, \tag{7.41}$$

$$D_4(k c_1 + c k_1) - k k_1 {D_2}^2 > 0. \tag{7.42}$$

Dividing inequalities given by (7.30)–(7.34), by the strictly positive quantity ${r_m}^2 \sin^2\lambda(1 + \mu\,\mathrm{sgn}(R\omega)\tan\lambda)$ gives

$$\tilde{D}_1 = I(m + \tilde{m}_1) > 0, \tag{7.43}$$

$$\tilde{D}_2 = c(m + \tilde{m}_1) + c_c\left(I - \xi_0 \tan\lambda m {r_m}^2\right) > 0, \tag{7.44}$$

$$\tilde{D}_3 = k(m + \tilde{m}_1) + c c_c + k_c\left(I - \xi_0 \tan\lambda m {r_m}^2\right) > 0, \tag{7.45}$$

$$\tilde{D}_2 \tilde{D}_3 - (k c_c + c k_c)\tilde{D}_1 > 0, \tag{7.46}$$

$$\left\{\tilde{D}_2 \tilde{D}_3 - (k c_c + c k_c)\tilde{D}_1\right\}(k c_c + c k_c) - k k_c {\tilde{D}_2}^2 > 0, \tag{7.47}$$

where

$$\tilde{m}_1 = \left(\frac{1}{\cos^2\lambda(1 + \mu\,\mathrm{sgn}(R\Omega)\tan\lambda)} - 1\right) m$$

and (7.14)–(7.16) were used.

Note that for $\mu > \tan\lambda$ and $R\Omega > 0$, $\tilde{m}_1$ is a small negative quantity. For example, for $\mu = 0.218$ and $\lambda = 5.57°$, $\tilde{m}_1/m \approx -0.012$. The conditions given by

(7.38)–(7.42) are structurally almost identical to (7.43)–(7.47). This indicates that, in the two models, k_1 and c_1 have the same effect on the eigenvalues as k_c and c_c. The difference between the two sets of stability conditions lies in (7.38) and (7.43) i.e., the term $-mm_1\xi_0 {r_m}^2 \tan\lambda$. However, for small lead screw mass (m_1), this difference is negligible.

Lemma 7.7. *For the linear system described by (7.8), the initial instability boundary is defined by either*

$$I(m + m_1) - mm_1\xi_0 {r_m}^2 \tan\lambda = 0 \tag{7.48}$$

or

$$D_4(kc_1 + ck_1) - kk_1{D_2}^2 = 0, \tag{7.49}$$

where D_2 and D_4 are given by (7.39) and (7.41), respectively.

Proof. Identical to Lemma 7.2. ■

Remark 7.5. The Lemmas 7.3–7.6 for the stability of the lead screw model with compliant threads in Sect. 7.2.3 can be restated for the 2-DOF model with axially compliant supports by replacing k_1 and c_1 with k_c and c_c, respectively. □

Remark 7.6. Because of the special structure of the inertia matrix of the linear system (7.8), in addition to the flutter boundary, (7.49), there exist a secondary instability boundary defined by (7.48). This additional boundary corresponds to the *kinematic constraint instability* which is the subject of the next chapter. □

7.4 Further Observations on the Mode Coupling Instability

Although the linear complex eigenvalue analysis method is useful in establishing the local stability boundaries of the equilibrium point in the system's parameter space, it does not reveal any information regarding dynamic behavior of the system. Further investigations such as numerical simulations or nonlinear analysis methods may be utilized to study the amplitude and frequency of the resulting vibrations under the mode coupling instability mechanism.

In this section, through numerical simulation, the effects of various system parameters on the dynamic behavior of the lead screw drive are investigated. The 2-DOF model defined by (7.11) is used in this section. First, in Sect. 7.4.1, a series of examples are presented where the evolution of dominant frequency of vibrations is plotted as the contact stiffness is varied. In these examples, using bifurcation plots of the double-sided Poincaré maps, evolution of the amplitude of vibrations is also investigated. In Sect. 7.4.2, the role of the two damping parameters (e.g., lead screw support rotational damping, c, and linear contact damping, c_c) on the amplitude of vibration of the lead screw is investigated.

7.4.1 Frequency and Amplitude of Vibrations

Figure 7.9 shows the results of a series of simulations where the contact stiffness is varied from 10^4 to 10^6 N/m. The value of the other system parameters is given in Table 7.1 and in the figure caption. In these simulations, the initial conditions were chosen as; $u_1(0) = u_2(0) = \dot{u}_2(0) = 0$ and $\dot{u}_1(0) = -\Omega/2$. For each set of system parameters, the numerical simulations were carried out for 4 s. The results for the first 2 s were discarded to exclude the transients.

Figure 7.9c shows the variation of the real part of the eigenvalues of the linearized system. The steady-sliding equilibrium point is unstable due to mode coupling for approximately $2.1 \times 10^5 < k_c < 6.6 \times 10^5$. The dominant frequency of the steady-state (angular) vibration of the lead screw is drawn in Fig. 7.9b and compared with the eigenfrequencies. The frequency of vibrations starts between the two damped natural frequencies and as the contact stiffness is increased exceeds the damped natural frequency of the higher mode. Corresponding to the same range of the contact stiffness, Fig. 7.9a shows the two-sided Poincaré bifurcation diagram. The Poincaré section was taken as $\dot{y}_1 = 0$. The variation of the steady-state vibration can be inferred from this figure.

Figure 7.10 shows the projection of the system trajectories for $k_c = 5 \times 10^5$ N/m. The stick-slip vibration of the lead screw due to mode coupling is clearly shown in this figure.

Figure 7.11 shows similar simulation results as in Fig. 7.9. In these results, the rotational damping coefficient of the lead screw supports, c, is reduced to 4×10^{-4}N m s/rad. As shown in Fig. 7.11c, the origin is now unstable for $7.1 \times 10^4 < k_c < 7.2 \times 10^5$ which is expanded compared with the pervious case. The maximum amplitude of vibration is considerably higher in this case, as can be

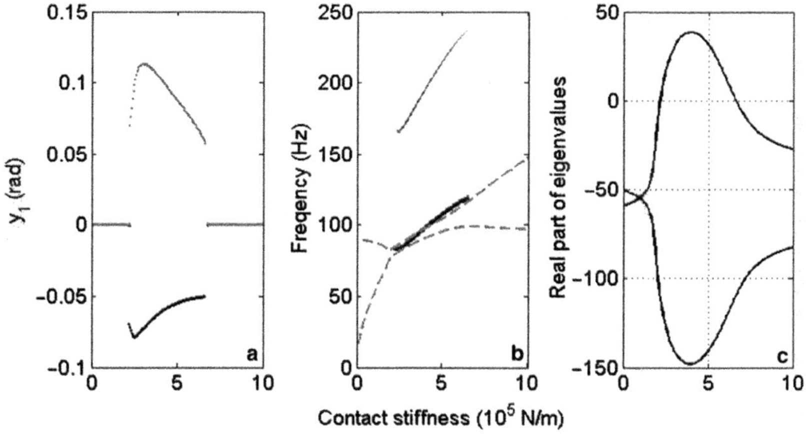

Fig. 7.9 Simulation results – effect of contact stiffness. $m = 1$, $c = 4 \times 10^{-4}$, and $c_c = 10^2$. (**a**) Two-sided Poincaré bifurcation diagram; (**b**) *Black lines*: frequency content of steady-state lead screw vibration, *dashed grey lines*: eigenfrequencies; (**c**) Real part of the eigenvalues

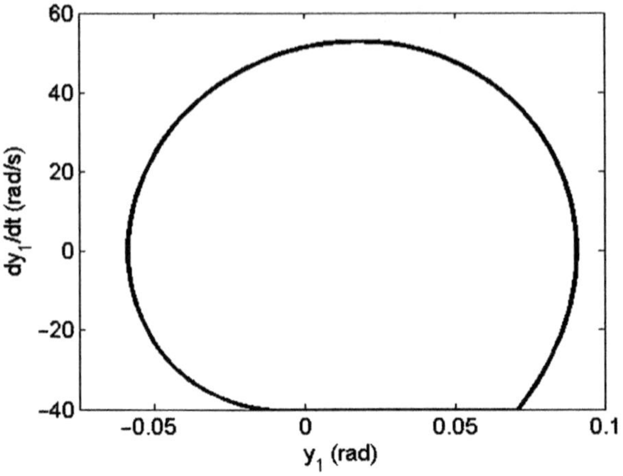

Fig. 7.10 Projection of the system trajectory on $y_1 - \dot{y}_1$ plane. $m = 1, c = 4 \times 10^{-4}, c_c = 10^2$, and $k_c = 5 \times 10^5$

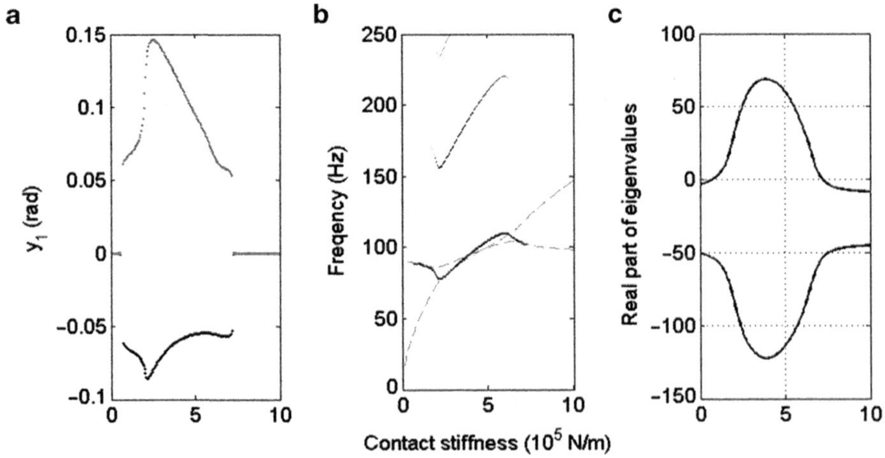

Fig. 7.11 Simulation results – effect of contact stiffness. $m = 1$, $c = 5 \times 10^{-5}$, and $c_c = 10^2$. (**a**) Two-sided Poincaré bifurcation diagram; (**b**) *Black lines*: frequency content of steady-state lead screw vibration, *dashed grey lines*: eigenfrequencies; (**c**) Real part of the eigenvalues

seen in the two-sided Poincare plot of Fig. 7.11a. It is interesting to note that the frequency of vibration first follows the upper mode as the two natural frequencies converge and then switches to the lower mode as the difference between the frequencies grows. The projection of the system trajectory on $y_1 - \dot{y}_1$ plane is shown in Fig. 7.12 for $k_c = 3 \times 10^5$.

The results for the final example in this series are shown in Fig. 7.13. In this example, compared with the first example above, both lead screw support damping

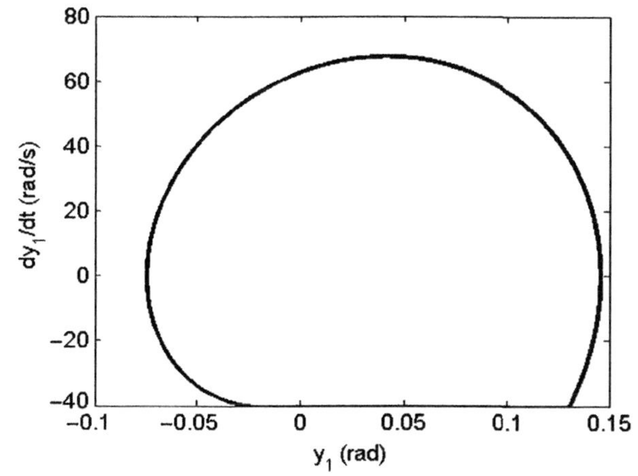

Fig. 7.12 Projection of the system trajectory on $y_1 - \dot{y}_1$ plane. $m = 1$, $c = 5 \times 10^{-5}$, $c_c = 10^2$, and $k_c = 3 \times 10^5$

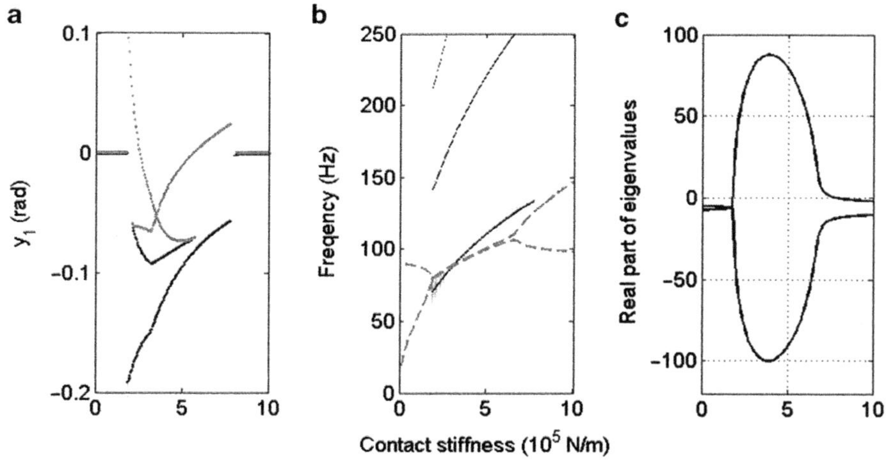

Fig. 7.13 Simulation results – effect of contact stiffness. $m = 1$, $c = 5 \times 10^{-5}$, and $c_c = 10$. (**a**) Two-sided Poincaré bifurcation diagram; (**b**) *Black lines*: frequency content of steady-state lead screw vibration, *dashed grey lines*: eigenfrequencies; (**c**) Real part of the eigenvalues

and contact damping coefficients are reduced; $c = 4 \times 10^{-4}\,\mathrm{N\,m\,s/rad}$ and $c_c = 10\,\mathrm{N\,s/m}$. The plot of the real part of the eigenvalues of the linearized system, Fig. 7.13c, shows that the instability region of the equilibrium point, has grown ($1.8 \times 10^5 < k_c < 7.8 \times 10^5$) compared with Fig. 7.9c. Interestingly, for the values $1.8 \times 10^5 < k_c < 5.7 \times 10^5$, the system exhibits period-doubled vibrations.

Sample limit cycle plots are shown in Fig. 7.14. The behavior is clearly seen for $k_c = 4 \times 10^5$ and $k_c = 2.5 \times 10^5\,\mathrm{N/m}$.

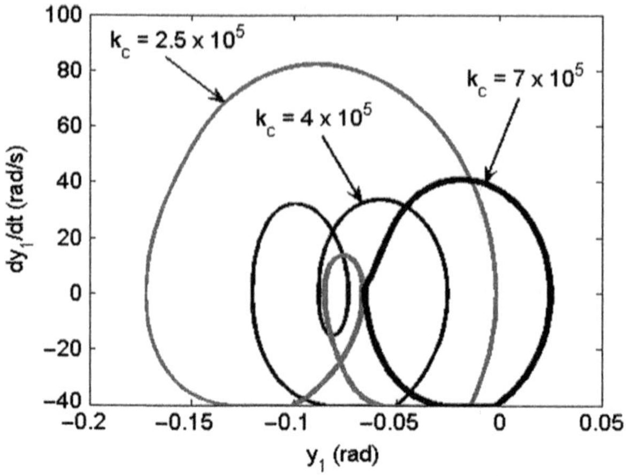

Fig. 7.14 Projection of the system trajectory on $y_1 - \dot{y}_1$ plane. $m = 1$, $c = 5 \times 10^{-5}$, and $c_c = 10$

At lower damping levels, the system may even exhibit chaos as we will see in the examples of the next section.

7.4.2 *Effect of Damping on Mode Coupling Vibrations*

Lemmas 7.5 and 7.6 showed that in the extreme cases where damping is present only in one of the two DOFs of the system, the steady-siding equilibrium point is unstable. In addition, the complex effect of damping in expanding *or* reducing the parameter regions of stability was shown by the examples in Sect. 7.2.4. The actual variations in the steady-state amplitude of vibrations can have an even more complex behavior.

Figure 7.15 shows a map of the averaged amplitude of vibrations of the lead screw for $k_c = 2 \times 10^7$, as lead screw support damping, and contact damping coefficients are varied. The other system parameters are given in Table 7.1 and $m = 5$ kg. The two natural frequencies of the undamped system are approximately 148.2 and 194.6 Hz. The initial conditions were chosen close to the equilibrium point; $y_1(0) = y_2(0) = 0$ and $\dot{y}_1(0) = \dot{y}_2(0) = -\Omega/10$. For each pair of damping coefficients, the numerical simulations were carried out for 4 s. The first second of results was discarded to exclude the transients. As can be seen, the steady-state amplitude of vibration varies considerably with the two damping coefficients.

For the numerical value of the parameters chosen here, the system exhibits chaotic or multiperiod behavior in many instances. Figure 7.16a, b shows the bifurcation diagrams of the Poincaré sections ($\dot{y}_1 = 0$), as the damping parameter is changed along the horizontal dotted line and the vertical dotted line in Fig. 7.15, respectively.

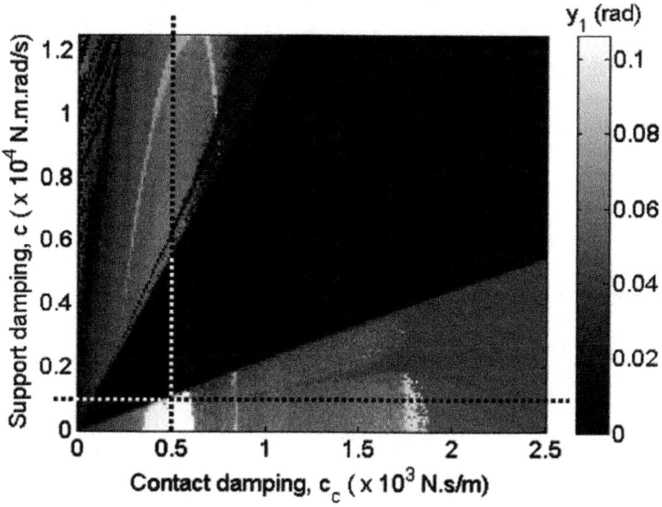

Fig. 7.15 Averaged amplitude of vibration, y_1 (rad), as lead screw support damping, c, and contact damping, c_c, are varied

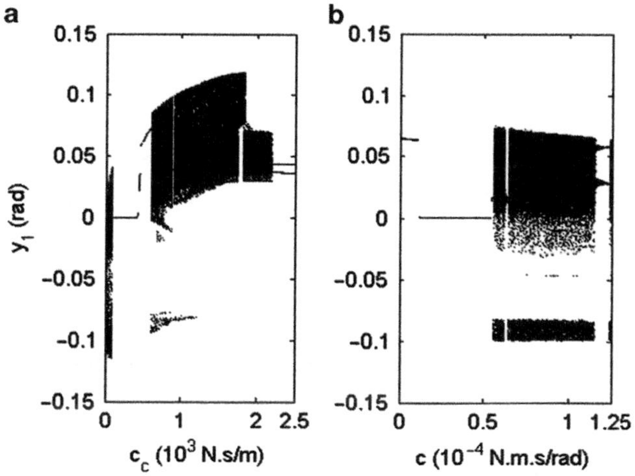

Fig. 7.16 Bifurcation of Poincaré sections. (**a**) Along the *horizontal dashed line* in Fig. 7.15; (**b**) along the *vertical dashed line* in Fig. 7.15

7.5 Mode Coupling in 3-DOF Lead Screw Model

In this section, local stability of the equilibrium point of the 3-DOF model described in Sect. 5.7 is investigated. Parameter studies and comparisons are done numerically. Similar to what was done in Sect. 7.1, the equations of motion are transformed to a suitable form corresponding to the steady-sliding equilibrium point. The equations of motion are given by (5.16), (5.17), and (5.28). Neglecting F_0, T_0, and c_x, these equations simplify to

$$I\ddot{\theta} = k(\theta_{\mathrm{i}} - \theta) - c\dot{\theta} + r_{\mathrm{m}}(N \sin\lambda - F_{\mathrm{f}} \cos\lambda), \tag{7.50}$$

$$m\ddot{x} = -N\cos\lambda - F_{\mathrm{f}}\sin\lambda + R, \tag{7.51}$$

$$m_1\ddot{x}_1 = -k_1x_1 - c_1\dot{x}_1 + N\cos\lambda + F_{\mathrm{f}}\sin\lambda. \tag{7.52}$$

Introducing the change of variables

$$\begin{aligned} u_1 &= \theta - \theta_{\mathrm{i}}, \\ u_2 &= x - r_{\mathrm{m}}\tan\lambda\theta_{\mathrm{i}}, \\ u_3 &= x_1, \end{aligned}$$

and setting all time derivatives to zero, the steady-sliding equilibrium point is found as

$$\begin{aligned} u_{10} &= -\frac{c\Omega + r_{\mathrm{m}}\xi_0 R}{k}, \\ u_{20} &= \frac{1}{k_{\mathrm{c}}\cos^2\lambda}\frac{R}{1+\mu\,\mathrm{sgn}(R\Omega)\tan\lambda} + r_{\mathrm{m}}\tan\lambda u_{10} + \frac{R}{k_1}, \\ u_{30} &= \frac{R}{k_1}. \end{aligned}$$

To transfer this point to the origin and present the system in state-space form, the following change of variables is applied:

$$\begin{aligned} y_1 &= \theta - \theta_{\mathrm{i}} - u_{10}, & y_4 &= \dot{y}_3, \\ y_2 &= \dot{y}_1, & y_5 &= x_1 - u_{30}, \\ y_3 &= x - r_{\mathrm{m}}\tan\lambda\theta_{\mathrm{i}} - u_{20}, & y_6 &= \dot{y}_5, \end{aligned}$$

which results in a system of six first-order differential equations

$$\dot{y}_i = f_i(\mathbf{y}), \; i = 1\ldots 6.$$

To study the stability of the steady-sliding equilibrium point, the eigenvalues of the Jacobian matrix (evaluated at $\mathbf{y} = 0$) are calculated. The Jacobian matrix is given by ($R \neq 0$, $\Omega \neq 0$)

$$\mathbf{A} = \left[\left.\frac{\partial f_i}{\partial y_j}\right|_{\mathbf{y}=0}\right] = \begin{bmatrix} 0 & 1 & 0 & 0 & 0 & 0 \\ \gamma_{21} & \gamma_{22} & \gamma_{23} & \gamma_{24} & \gamma_{25} & \gamma_{26} \\ 0 & 0 & 0 & 1 & 0 & 0 \\ \gamma_{41} & \gamma_{42} & \gamma_{43} & \gamma_{44} & \gamma_{45} & \gamma_{46} \\ 0 & 0 & 0 & 0 & 0 & 1 \\ \gamma_{61} & \gamma_{62} & \gamma_{63} & \gamma_{64} & \gamma_{65} & \gamma_{66} \end{bmatrix},$$

Table 7.2 Elements of the Jacobian matrix for the 3-DOF model

$\gamma_{21} = -\frac{k}{I} + \frac{\rho_{10} k_c r_m^2 \sin\lambda}{I}$	$\gamma_{41} = \frac{\rho_{20} k_c r_m \sin\lambda}{m}$	$\gamma_{61} = -\frac{\rho_{20} k_c r_m \sin\lambda}{m_1}$
$\gamma_{22} = -\frac{c}{I} + \frac{\rho_{10} c_c r_m^2 \sin\lambda}{I}$	$\gamma_{42} = \frac{\rho_{20} c_c r_m \sin\lambda}{m}$	$\gamma_{62} = -\frac{\rho_{20} c_c r_m \sin\lambda}{m_1}$
$\gamma_{23} = -\frac{r_m \rho_{10} k_c \cos\lambda}{I}$	$\gamma_{43} = -\frac{\rho_{20} k_c \cos\lambda}{m}$	$\gamma_{63} = \frac{\rho_{20} k_c \cos\lambda}{m_1}$
$\gamma_{24} = -\frac{r_m \rho_{10} c_c \cos\lambda}{I}$	$\gamma_{44} = -\frac{\rho_{20} c_c \cos\lambda}{m}$	$\gamma_{64} = \frac{\rho_{20} c_c \cos\lambda}{m_1}$
$\gamma_{25} = \frac{r_m \rho_{10} k_c \cos\lambda}{I}$	$\gamma_{45} = \frac{\rho_{20} k_c \cos\lambda}{m}$	$\gamma_{65} = -\frac{k_1}{m_1} - \frac{\rho_{20} k_c \cos\lambda}{m_1}$
$\gamma_{26} = \frac{r_m \rho_{10} c_c \cos\lambda}{I}$	$\gamma_{46} = \frac{\rho_{20} c_c \cos\lambda}{m}$	$\gamma_{66} = -\frac{c_1}{m_1} - \frac{\rho_{20} c_c \cos\lambda}{m_1}$
$\rho_{10} = -\sin\lambda + \mu_0 \mathrm{sgn}(R\Omega)\cos\lambda$		$\rho_{20} = \cos\lambda + \mu_0 \mathrm{sgn}(R\Omega)\sin\lambda$

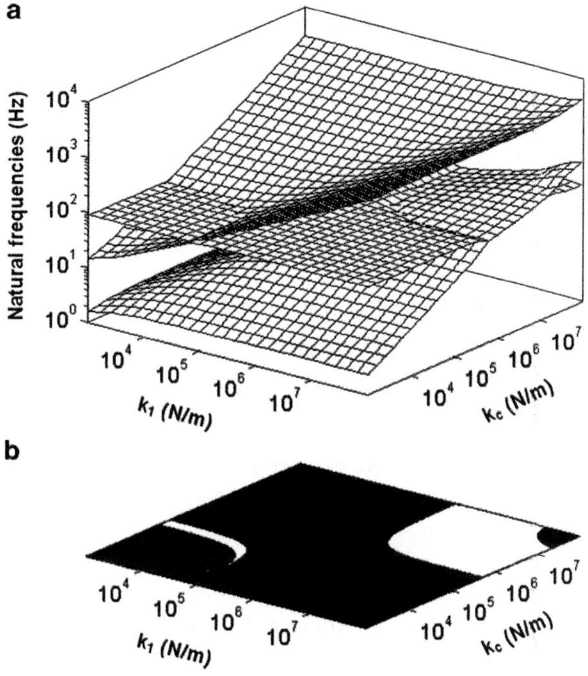

Fig. 7.17 (**a**) Evolution of the three natural frequencies of the undamped 3-DOF system (with constant coefficient of friction) as a function of k_c and k_1. (**b**) Stability map

where γ_{ij}s are given in Table 7.2.

Figure 7.17a shows the variation of the three undamped natural frequencies of the 3-DOF model with constant coefficient of friction as a function of lead screw

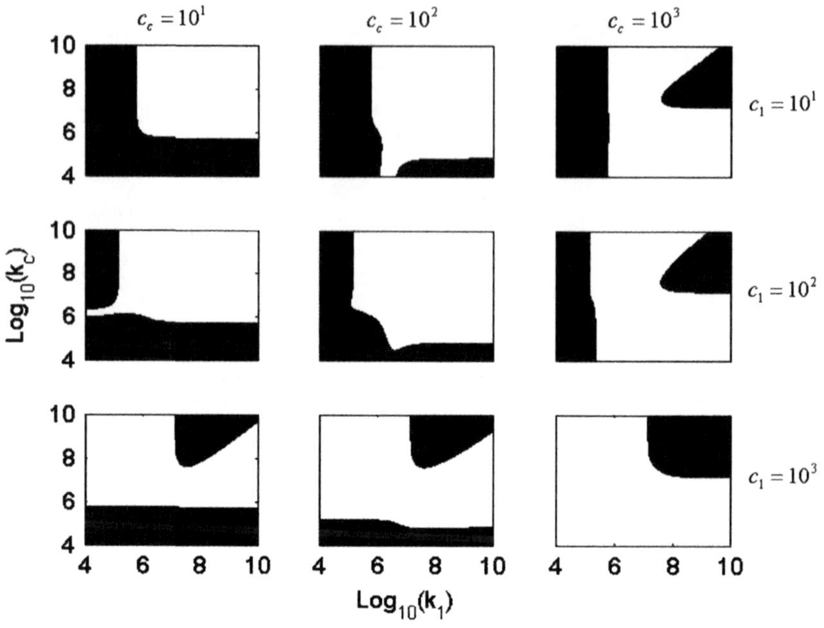

Fig. 7.18 Local stability of equilibrium points of the 3-DOF lead screw system with constant coefficient of friction. *Black*: stable, *white*: unstable

support stiffness (k_1) and contact stiffness (k_c). Lead screw parameters are those listed in Table 7.1 together with $m = 5$ kg and $m_1 = 0.232$ kg. The corresponding stability map, which is obtained by examining the real part of the eigenvalues, is depicted in Fig. 7.17b. This map shows that the origin becomes unstable whenever two of the system modes merge. Note that the parameter range where coupling between the first and the second modes occurs agrees with the instability range of the undamped 2-DOF model of Sect. 7.2.1 for large values of k_1.

Figure 7.18 shows the stability maps of the 3-DOF model as the contact stiffness (k_c) and the support stiffness (k_1) are varied. In the 3 by 3 series of plots included in this figure, the contact damping coefficient (c_c) and lead screw support translational damping coefficient (c_1) take the values: 10, 10^2, and 10^3 N s/m. Also the lead screw damping (angular) coefficient is chosen as $c = 3 \times 10^{-4}$ N m s/rad. Other parameters are selected as before. These plots clearly show the role of damping in both *stabilizing* an unstable equilibrium point and *destabilizing* a stable one. From the symmetry of the plots and for the selected values and ranges of values of parameters, one can deduce that the stiffness and damping of the two translational DOFs (i.e., x and x_1) have similar effects on the stability of the system. This, of course, agrees with our conclusions in Sect. 7.3.

7.6 Conclusions

In this chapter, the mode coupling instability in the lead screw drives was studied using several multi-DOF models. It was found that the necessary conditions for the mode coupling instability to occur are: (a) the lead screw must be self-locking (i.e., $\mu > \tan\lambda$) and (b) the direction of the applied axial force must be the same as the direction of motion of the translating part (i.e., $R\Omega > 0$). The flutter instability boundary in the space of system parameters for the 2-DOF models of Sects. 5.5 and 5.6 was given by (7.36) and (7.49), respectively.

As shown by the numerical simulation results of Sect. 7.4, mode coupling instability mechanism can lead to diverse range of system behaviors: from simple stick-slip limit cycles to complex multiperiod or chaotic responses.

In this chapter, using a 3-DOF model, it was shown that when mode coupling instability mechanism can affect a system, all the relevant DOFs must be included in the model. It was also shown that the compliance caused by the thread flexibility has similar effects on the stability of the system as the axial compliance in the lead screw supports.

Chapter 8
Kinematic Constraint Instability Mechanism

The third and final instability mechanism in the lead screw drives is the kinematic constraint. In Sect. 4.3, Painlevé's paradoxes were introduced and – through simple examples – it was shown that under the conditions of the paradoxes, the rigid body equations of motion of a system with frictional contact do not have a bounded solution or the solution is not unique. We have also discussed the relationship between Painlevé's paradoxes and the kinematic constraint instability mechanism.

We start this chapter by the investigation of the possibility of the occurrence of Painlevé's paradoxes in lead screw drives in Sect. 8.1. A discussion regarding the true motion of the system when paradoxes occur is presented in Sect. 8.2. The 1-DOF dynamic model of a lead screw drive developed in Sect. 5.3 is chosen for the study of the instabilities caused by the kinematic constraint mechanism. The equation of motion of this model is revisited in Sect. 8.3. Using eigenvalue analysis, the stability of the steady-sliding equilibrium point of the 1-DOF model is investigated in Sect. 8.4. We will show that the system loses stability in the region of paradoxes. The negative damping effect of the kinematic constraint is discussed in Sect. 8.5. The instabilities caused by the paradoxes (inertia effects) are studied in Sect. 8.6. A short discussion regarding the region of attraction of the stable steady-sliding equilibrium point is given in Sect. 8.7. We will see that even when the equilibrium point is locally stable, sufficiently large perturbations can lead to periodic vibrations. Kinematic constraint instability mechanism in multi-DOF models of lead screw drives is the topic of Sect. 8.8. Finally, major findings of this chapter are summarized in Sect. 8.9.

8.1 Existence and Uniqueness Problem

Consider the interacting lead screw and nut pair shown in Fig. 5.3. The equation of motion of this system is given by (5.13):

$$\Gamma\ddot{\theta} = T - r_{\mathrm{m}}\xi P, \tag{8.1}$$

O. Vahid-Araghi and F. Golnaraghi, *Friction-Induced Vibration in Lead Screw Drives*,
DOI 10.1007/978-1-4419-1752-2_8, © Springer Science+Business Media, LLC 2011

where Γ is the "effective inertia" and is given by:

$$\Gamma = I - \xi \tan \lambda m r_{\mathrm{m}}^2 \tag{8.2}$$

and ξ is defined by (5.11):

$$\xi = \frac{\mu_{\mathrm{s}} - \tan \lambda}{1 + \mu_{\mathrm{s}} \tan \lambda},$$

where the abbreviation $\mu_{\mathrm{s}} = \mu \operatorname{sgn}(\dot{\theta}) \operatorname{sgn}(N)$ was used. Also, the normal contact force is given by (5.14):

$$N = \frac{PI - m r_{\mathrm{m}} \tan \lambda T}{\Gamma(\cos \lambda + \mu_{\mathrm{s}} \sin \lambda)}. \tag{8.3}$$

Now we consider the possibility of Painlevé's paradoxes for the equation of motion of the lead screw drive given by (8.1). Define

$$\xi = \begin{cases} \xi_+ = \dfrac{\mu - \tan \lambda}{1 + \mu \tan \lambda}, & \operatorname{sgn}(\dot{\theta}) \operatorname{sgn}(N) = 1 \\ \xi_- = -\dfrac{\mu + \tan \lambda}{1 - \mu \tan \lambda}, & \operatorname{sgn}(\dot{\theta}) \operatorname{sgn}(N) = -1 \end{cases}$$

and

$$\Gamma = \begin{cases} \Gamma_+ = I - m r_{\mathrm{m}}^2 \tan \lambda \xi_+, & \operatorname{sgn}(\dot{\theta}) \operatorname{sgn}(N) = 1 \\ \Gamma_- = I - m r_{\mathrm{m}}^2 \tan \lambda \xi_-, & \operatorname{sgn}(\dot{\theta}) \operatorname{sgn}(N) = -1 \end{cases}. \tag{8.4}$$

Note that we have $\xi_- < 0$[1] and $\Gamma_- > 0$. Assume that the system's parameters are selected such that $\Gamma_+ < 0$ which requires that $\xi_+ > 0$ and $m r_{\mathrm{m}}^2 \tan \lambda \xi_+ > I$. For ξ_+ to be positive, coefficient of friction must be large enough such that $\mu > \tan \lambda$.

If $\operatorname{sgn}(\dot{\theta}) \operatorname{sgn}(PI - m r_{\mathrm{m}} \tan \lambda T) = 1$, then (8.3) has *no solution*; setting $\operatorname{sgn}(N) = 1$ in the RHS of (8.3) results in a negative contact force, and setting $\operatorname{sgn}(N) = -1$ results in a positive contact force. On the other hand, if $\operatorname{sgn}(\dot{\theta}) \operatorname{sgn}(PI - m r_{\mathrm{m}} \tan \lambda T) = -1$, then (8.3) has *two distinct solutions:*

$$N = \begin{cases} N_+ = \dfrac{PI - m r_{\mathrm{m}} \tan \lambda T}{(\cos \lambda + \mu \sin \lambda)\Gamma_+}, & \operatorname{sgn}(\dot{\theta}) \operatorname{sgn}(N) = 1, \\ N_- = \dfrac{PI - m r_{\mathrm{m}} \tan \lambda T}{(\cos \lambda - \mu \sin \lambda)\Gamma_-}, & \operatorname{sgn}(\dot{\theta}) \operatorname{sgn}(N) = -1. \end{cases} \tag{8.5}$$

[1]See footnote on page 113.

In the next section, we use the *limiting process* approach [51] to determine the true motion of the system in the regions of the paradoxes.

8.2 True Motion in Paradoxical Situations

In Sect. 4.3.4.4 we studied an example where an approximate solution was obtained in the region of paradoxes by adding compliance to the two bodies in contact. In Sect. 8.6.1 below, we will present numerical results of a similar approach applied to the lead screw and nut (i.e. using the 2-DOF model of Sect. 5.5). But first, we will take a closer look at the behavior of the rigid body system under the conditions of the paradoxes. The approach adopted here is based on the limiting process described in [51] where the law of motion of the rigid body system is taken as that of the system with compliant contact when the contact stiffness tends to infinity.

The equations of motion of lead screw and nut with complaint threads are given by (5.16) and (5.17). Instead of (5.24), the contact force is assumed here as (neglecting contact damping):

$$N = k_c \delta, \tag{8.6}$$

where δ is given by (5.23). Differentiating (8.6) twice with respect to time and using (5.23), (5.16), and (5.17), yield

$$\ddot{N} = \frac{k_c \cos\lambda}{mI}[-(1+\mu_s \tan\lambda)\cos\lambda \Gamma N + (IP - r_m \tan\lambda mT)], \tag{8.7}$$

which is a second-order differential equations with constant coefficients. Note that outside the region of paradoxes, the stationary solution of (8.7) coincides with (8.3). Let

$$\chi = \frac{N}{\hat{N}}, \tag{8.8}$$

where

$$\hat{N} = \frac{PI - mr_m \tan\lambda T}{(\cos\lambda + \mu_s \sin\lambda)}.$$

In addition, by using the nondimensional time τ

$$\tau = \bar{\omega} t, \tag{8.9}$$

where

$$\bar{\omega} = \sqrt{\frac{k_c \cos\lambda}{mI}(1+\mu_s \tan\lambda)\cos\lambda|\Gamma|} \tag{8.7}$$

(8.7) is transformed to:

$$\ddot{\chi} + \frac{\Gamma}{|\Gamma|}\chi - \frac{1}{|\Gamma|} = 0. \tag{8.10}$$

Now consider the case where conditions given by (8.21) are satisfied. The differential equation given by (8.10) can be written as:

$$\begin{cases} \ddot{\chi} - \chi + \Gamma_+{}^{-1} = 0, & N\dot{\theta} > 0 \\ \ddot{\chi} + \chi - \Gamma_-{}^{-1} = 0, & N\dot{\theta} < 0 \end{cases}. \tag{8.11}$$

The fixed point of the first equation in (8.11) is a saddle at $\chi = \Gamma_+{}^{-1}$ and second equation has a center at $\chi = \Gamma_-{}^{-1}$. The solutions of the these equations can be written as

$$\begin{cases} \chi(\tau) = a_1 e^{\tau} + a_2 e^{-\tau} + \Gamma_+{}^{-1}, & N\dot{\theta} > 0 \\ \chi(\tau) = b_1 \sin\tau + b_2 \cos\tau + \Gamma_-{}^{-1}, & N\dot{\theta} < 0 \end{cases}. \tag{8.12}$$

First let's consider the indeterminacy paradox. As mentioned in Sect. 8.1, a necessary condition for indeterminacy is $\text{sgn}(\dot{\theta})\text{sgn}(PI - mr_{\text{m}} \tan \lambda T) = -1$. For simplicity we only consider here the case of $\dot{\theta} > 0$. The phase portrait of the differential equation (8.11) – for this case – is shown in Fig. 8.1. It is important to notice that the contact force for any initial condition is uniquely defined by (8.11).

As can be seen from Fig. 8.1, two outcomes are possible depending on the initial conditions: The contact force remains near the center $\chi = \Gamma_-{}^{-1}$ or the contact force diverges to negative infinity exponentially.

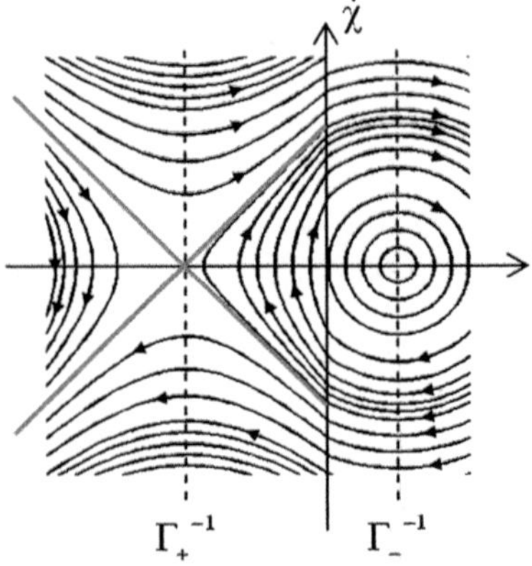

Fig. 8.1 Phase portrait of the differential equation for the contact force – indeterminate case

Remark 8.1. If damping force $(c_c\dot{\delta})$ is added to contact force (8.6), (8.10) becomes

$$\ddot{\chi} + \frac{\Gamma}{|\Gamma|}\left(\chi + \frac{c_c}{k_c}\dot{\chi}\right) - \frac{1}{|\Gamma|} = 0.$$

Consequently, the second equation of (8.11) becomes: $\ddot{\chi} + (c_c/k_c)\dot{\chi} + \chi - {\Gamma_-}^{-1} = 0$ which has a stable focus at $\chi = {\Gamma_-}^{-1}$. Using (8.8), the steady-state contact force is found to be $N = N_-$ where N_- is given by (8.5). □

For the case where the contact force grows without bound, one can show that as $k_c \to 0$, a discontinuous change occurs in the velocity which is known as the *tangential impact* and motion stops instantaneously. Substituting the first equation of (8.12) into (5.16) gives

$$I\ddot{\theta} = T - r_m\xi_+(PI - mr_m \tan\lambda T)(a_1 e^{\varpi t} + a_2 e^{-\varpi t} + {\Gamma_+}^{-1}), \quad N\dot{\theta} > 0,$$

where (8.9) was used. The changes in the velocity for a time interval Δt is

$$\begin{aligned}\Delta\dot{\theta} &= \int_0^{\Delta t} \ddot{\theta}\mathrm{d}t, \\ &= TI^{-1}\Delta t - r_m\xi_+\left(P - mr_m \tan\lambda TI^{-1}\right) \\ &\quad \times \left(a_1\varpi^{-1}\left(e^{\varpi\Delta t} - 1\right) - \varpi^{-1}a_2\left(e^{-\varpi\Delta t} - 1\right) + {\Gamma_+}^{-1}\Delta t\right).\end{aligned}$$

Taking the limit of this expression as $k_c \to 0$, gives

$$\begin{aligned}\lim_{k_c\to\infty} \Delta\dot{\theta} &= \lim_{k_c\to\infty} \left[TI^{-1}\Delta t - r_m\xi_+(P - mr_m \tan\lambda TI^{-1})(a_1\varpi^{-1}(e^{\varpi\Delta t} - 1)\right. \\ &\quad \left. -\varpi^{-1}a_2(e^{-\varpi\Delta t} - 1) + {\Gamma_+}^{-1}\Delta t)\right], \\ &= \lim_{k_c\to\infty}\left[-r_m\frac{\mu - \tan\lambda}{1 + \mu\tan\lambda}\left(P - mr_m \tan\lambda TI^{-1}\right)a_1\varpi^{-1}\left(e^{\varpi\Delta t} - 1\right)\right], \\ &= \kappa \lim_{k_c\to\infty}\left[\varpi^{-1}e^{\varpi\Delta t}\right],\end{aligned}$$

where $\kappa = -r_m((\mu - \tan\lambda)/(1 + \mu\tan\lambda))(P - mr_m \tan\lambda TI^{-1})a_1$. The last equality can be interpreted as follows: For a bounded change in velocity $\Delta\dot{\theta}$, the time interval Δt required tends to zero as $k_c \to 0$. Thus, velocity becomes discontinuous in the limit as the contact stiffness tends to infinity.

Now consider the inconsistency situation; i.e., $\mathrm{sgn}(\dot{\theta})\mathrm{sgn}(PI - mr_m \tan\lambda T) = 1$. The phase portrait of the differential equation (8.11) is shown in Fig. 8.2. Although the differential equation (8.10) *does not* have an equilibrium point, similar to the indeterminate case, the contact force is uniquely defined.

Figure 8.2 shows that, for any initial condition, the contact force eventually reaches the first quadrant and grows exponentially without bound. Similar to the previous case, one can show that in the limit as $k_c \to 0$ velocity changes discontinuously and a tangential impact occurs.

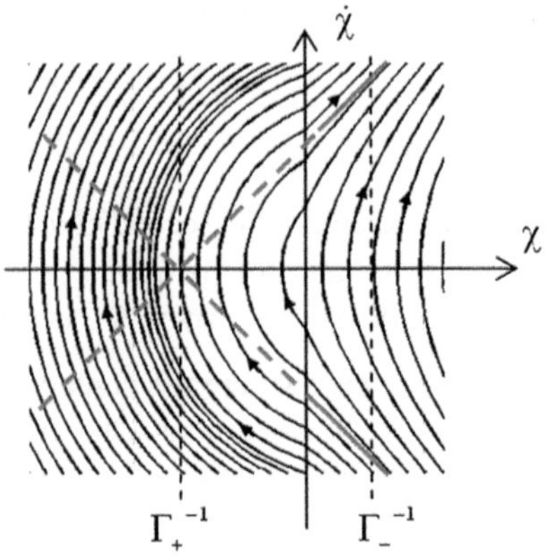

Fig. 8.2 Phase portrait of the differential equation for the contact force – inconsistent case

In [51] (corollary to theorems 8 and 9), it is proven that whenever tangential impact occurs (in both the indeterminate and the inconsistent cases) dynamic seizure occurs and motion stops instantaneously.

8.3 1-DOF Lead Screw Drive Model

Consider the 1-DOF lead screw drive model in Fig. 5.4. Here, for simplicity, the Coulomb friction of the translating part is neglected (i.e., $F_0 = 0$). The equation of motion is given by (5.18):

$$\Gamma\ddot{\theta} + k\theta + C\dot{\theta} = k\theta_{\mathrm{i}} - r_{\mathrm{m}}\xi R, \tag{8.13}$$

where the abbreviations (8.2) and

$$C = c - \xi r_{\mathrm{m}}^2 \tan\lambda c_{\mathrm{x}}$$

were used. Let $z = \theta - \theta_{\mathrm{i}}$ then $\dot{z} = \dot{\theta} - \Omega$ and $\ddot{z} = \ddot{\theta}$, where $\Omega = \mathrm{d}\theta_{\mathrm{i}}/\mathrm{d}t$ is a nonzero constant representing the input angular velocity. Substituting this change of variable into (8.13) gives

$$\Gamma\ddot{z} + C\dot{z} + kz = -C\Omega - r_{\mathrm{m}}\xi R. \tag{8.14}$$

At steady-sliding we have $\ddot{z} = 0$, $\dot{z} = 0$, and $z = z_0$. Substituting these values in (8.14) yields

$$z_0 = -\frac{C_0\Omega + r_{\mathrm{m}}\xi_0 R}{k},$$

where $C_0 = c - \xi_0 r_m^2 \tan\lambda c_1$ and $\xi_0 = (\mu \text{sgn}(N_0\Omega) - \tan\lambda)/(1 + \mu \text{sgn}(N_0\Omega)\tan\lambda)$ where

$$N_0 = \frac{R - r_m \tan\lambda c_1 \Omega}{\cos\lambda + \mu \text{sgn}(N_0\Omega)\sin\lambda}. \tag{8.15}$$

Note that for the applicable range of $0 \le \mu < \cot\lambda$, we have $\text{sgn}(N_0) = \text{sgn}(R - r_m \tan\lambda c_1\Omega)$. Using the change of variable $u = z - z_0$, the steady-sliding state is transferred to the origin and (8.14) becomes

$$\Gamma\ddot{u} + C\dot{u} + ku = r_m(\xi_0 - \xi)(R - r_m \tan\lambda c_1 \Omega). \tag{8.16}$$

In terms of the new variable, the contact force is given by

$$N = \frac{r_m \tan\lambda[mku + (mc - Ic_1)\dot{u}] + \Gamma_0(R - c_1 r_m \tan\lambda\Omega)}{\Gamma(\cos\lambda + \mu_s \sin\lambda)}, \tag{8.17}$$

where $\Gamma_0 = I - \tan\lambda\xi_0 m r_m^2$.

8.4 Linear Stability Analysis

Using the variables, $y_1 = u$ and $y_2 = \dot{u}$, (8.16) is converted to a system of first-order differential equations. The Jacobian matrix of this system evaluated at the origin is calculated as

$$A = \begin{bmatrix} 0 & 1 \\ -\frac{k}{\Gamma_0} & -\frac{C_0}{\Gamma_0} \end{bmatrix}, \quad \Gamma_0 \neq 0, N_0 \neq 0, \Omega \neq 0.$$

The eigenvalues of the Jacobian matrix are

$$e_1, e_2 = -\frac{C_0}{2\Gamma_0} \pm \frac{1}{2|\Gamma_0|}\sqrt{C_0{}^2 - 4k\Gamma_0} \tag{8.18}$$

Figure 8.3 shows the region of stability of the steady-sliding equilibrium point as Γ_0 and C_0 are varied. Due to the presence of the kinematic constraint in the system and for a self-locking lead screw drive (i.e., $\mu > \tan\lambda$), both of these parameters can attain negative values. The origin of the linearized system is unstable whenever $\Gamma_0 < 0$ or $C_0 < 0$.

For the parameter values where $\Gamma_0 < 0$, however, we know that the original equation of motion of the system is no longer valid due to the occurrence of paradoxes. Using the limiting process approach, the following lemma proves that indeed the 1-DOF lead screw system loses stability when $\Gamma_0 < 0$.

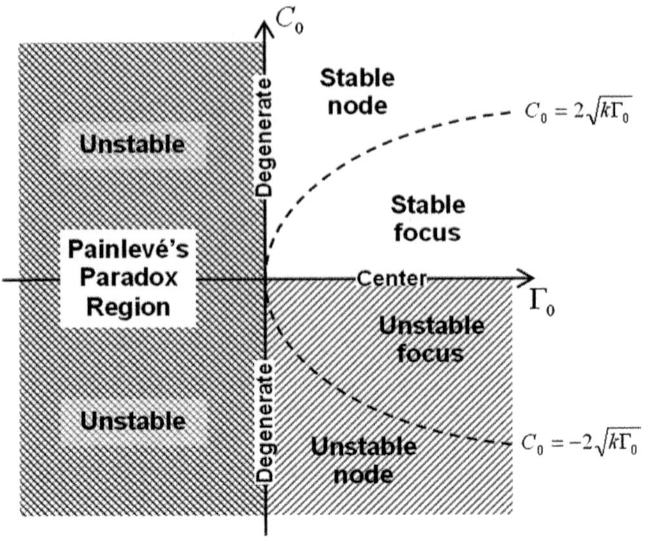

Fig. 8.3 Stability/instability regions of the steady-sliding equilibrium point as Γ_0 and C_0 are varied

Table 8.1 Parameters values used in the numerical simulations

Parameter	Value	Unit
r_m	5.18	mm
λ	5.57	degree
I	3.12×10^{-6}	kg m^2
k	1	N m/rad
c	2×10^{-4}	N m s/rad
μ	0.218	–
Ω	40	rad/s

Lemma 8.1:. *The origin is an unstable equilibrium point for system (8.16) whenever* $\Gamma_0 < 0$.

Proof. For simplicity, we assume $c_1 = 0$. Consider the system in Fig. 5.9 with $F_0 = 0$ and $T_0 = 0$. The linear stability conditions for this system are those given by (7.30) to (7.34). First, notice that $\text{sgn}(\Gamma_0) = \text{sgn}(\Delta)$, where Δ is given by (7.35). As $k_c \to \infty$, based on (7.32), the origin is unstable if $\Delta < 0$. ■

It will be shown by numerical examples in Sect. 8.5 below that when $\Gamma_0 > 0$ and $C_0 < 0$ (i.e., negative effective damping), the instability may or may not lead to stick-slip vibrations. In contrast to this case, when $\Gamma_0 < 0$ (i.e., Painlevé's paradox), the instability is accompanied by stick-slip vibration and impulsive forces. Section 8.6 is dedicated to the study of the kinematic constraint instability and the resulting vibrations.

In the subsequent sections and unless otherwise specified the values of system parameters used in numerical simulations are those given in Table 8.1.

8.5 Negative Damping

In Chap. 6, we saw that a decreasing coefficient of friction with relative sliding velocity may lead to instability. Here, a different mechanism is discussed where negative damping may exist even with constant coefficient of friction. This type of negative damping is a direct consequence of the kinematic constraint relationship describing the lead screw with friction. The necessary and sufficient conditions for the *effective damping*, C, to be negative are

$$\begin{aligned} \mu &> \tan\lambda, \\ N_0\Omega &> 0, \\ \xi_+ {r_m}^2 \tan\lambda c_1 &> c, \end{aligned} \tag{8.19}$$

where N_0 is given by (8.15). Under these conditions and assuming $\Gamma > 0$ for $\forall u, \dot{u} \in R$, the steady-sliding equilibrium point is unstable. Fig. 8.4 shows the system's phase plane when conditions of (8.19) are satisfied and $\Omega > 0$. Trajectories on the $N > 0$ half-plane are governed by the (unstable) differential equation:

$$\Gamma_+\ddot{u} + C_+\dot{u} + ku = 0,$$

where Γ_+ is defined by (8.4) and it is assumed that $\Gamma_+ > 0$. Also,

$$C_+ = c - \xi_+ r_m^2 \tan\lambda c_x$$

and according the (8.19) we have $C_+ < 0$. Trajectories on the $N < 0$ half-plane are solutions of the (stable) differential equation:

$$\Gamma_-\ddot{u} + C_-\dot{u} + ku = r_m(\xi_+ - \xi_-)(R - r_m \tan\lambda c_x\Omega),$$

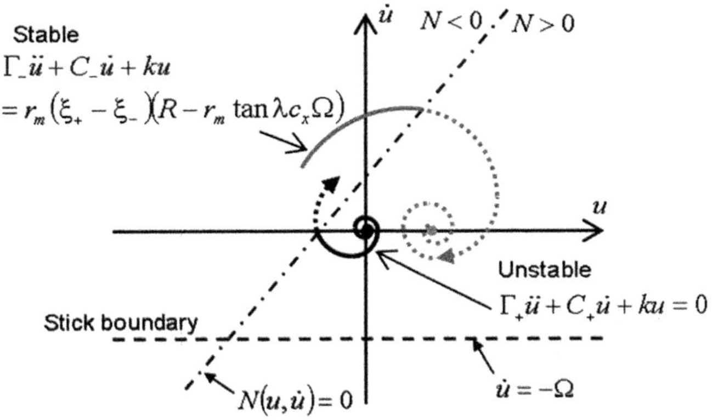

Fig. 8.4 Unstable steady sliding equilibrium point due to damping kinematic constraint instability mechanism $\mu > \tan\lambda$, $N_0 > 0$, $\Omega > 0$, and $\xi_+ r_m^2 \tan\lambda c_x > c$

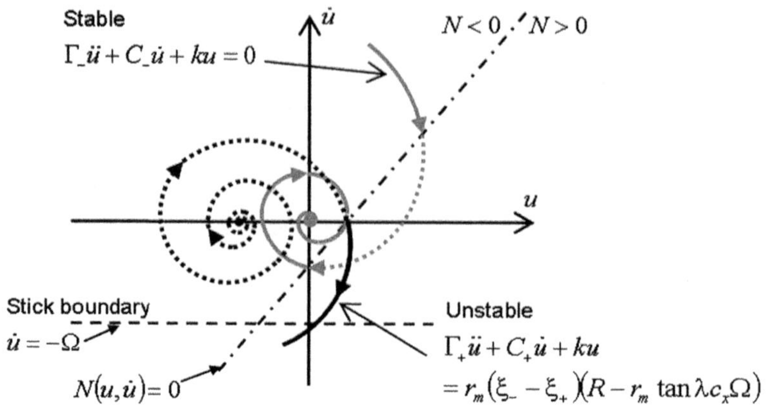

Fig. 8.5 Stable steady-sliding equilibrium point $\mu > \tan\lambda$, $N_0 < 0$, $\Omega > 0$, and $\xi_{+}r_{\rm m}^2\tan\lambda c_{\rm x} > c$

where $\Gamma_{-} > 0$ is defined by (8.4). Also,

$$C_{-} = c - \xi_{-}r_{\rm m}^2 \tan\lambda c_{\rm x} > 0.$$

By decreasing R (the axial force) the line $N = 0$ in Fig. 8.4 moves towards the origin. Fig. 8.5 shows a situation where the steady-sliding equilibrium point is in the $N < 0$ half plane.

Assuming other system parameters are unchanged, the second inequality of the conditions given by (8.19) is now violated (i.e., $N_0\Omega < 0$) and consequently the equilibrium point is stable.

8.5.1 *Numerical Simulation Results*

In the simulation results shown in Fig. 8.6, variation of the steady-state amplitude of vibration vs. the axial force is plotted. In these simulations, parameter values are taken from Table 8.1. In addition, the mass and the damping coefficient of the translating part are set to $m = 1$ kg and $c_1 = 10^3$N s/m. First, notice that the self-locking condition is satisfied for the selected value of the constant coefficient of friction (i.e., $\mu = 0.218 > \tan(5.57°) = 0.0975$). Also note that, using the selected parameters;

$$C_{-} \approx 1.0 \times 10^{-3} N\,s/m > 0, \quad \Gamma_{-} \approx 9.9 \times 10^{-6}\,kg\,m^2 > 0,$$

$$C_{+} \approx -1.1 \times 10^{-4} N\,s/m < 0, \quad \Gamma_{+} \approx 6.5 \times 10^{-7}\,kg\,m^2 > 0.$$

As shown in Fig. 8.6, at low values of applied axial force (approximately $R < 20\,N$), the origin is stable since $R - r_{\rm m}\tan\lambda c_{\rm x}\Omega < 0$, which means that $N_0 < 0$

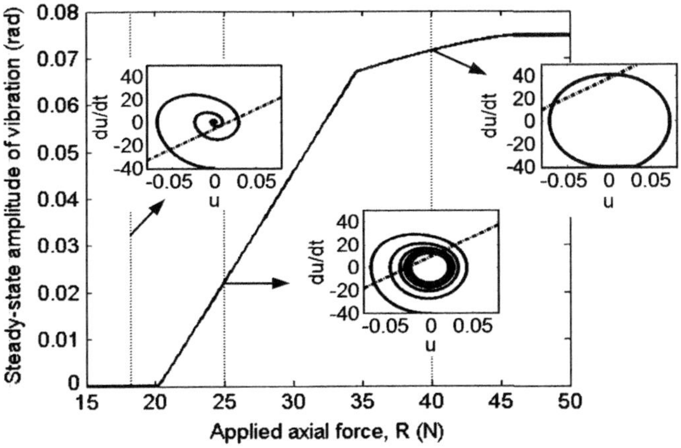

Fig. 8.6 Steady-state vibration amplitude as a function of applied axial force, R. $c_1 = 10^3 \mathrm{N\,s/m}$, $m = 1$ kg

and the conditions of (8.19) are not satisfied. However, once that applied axial force is such that $N_0 > 0$, the steady-sliding equilibrium point is unstable and periodic vibrations develop.

For the approximate range of $20 < R < 34.5N$, the resulting limit cycle does not touch the stick-slip boundary (i.e., $\dot{u} = -\Omega$) and the amplitude of vibrations grow almost linearly with the increase of R. For forces greater than approximately $34.5N$, stick-slip vibrations are observed. Beyond approximately $46.5N$, the limit cycles does not intersect the $N = 0$ line and the amplitude of steady-state vibrations remains constant. The three inset plots in Fig. 8.6 show the phase trajectories for three values of applied axial force. At $R = 18N$, the steady-sliding equilibrium point is stable at trajectories asymptotically reach the origin. At $R = 25N$ the steady-sliding equilibrium point is unstable and the trajectories are attracted to a pure-slip limit cycle. In this case, the periodic vibrations do not have a "sticking" phase. At $R = 40N$ the increase of the applied axial force has expanded the limit cycle, which now touches the stick-slip boundary. The periodic vibrations in this case resemble a typical stick-slip vibration in systems with decreasing coefficient of friction with relative sliding velocity.

8.6 Kinematic Constraint Instability

From the eigenvalue analysis of Sect. 8.4, it is evident that regardless of the level of linear damping (C_0), instability (due to Painlevé's paradoxes) occurs whenever

$$\Gamma_0 = I - \xi_0 m {r_m}^2 \tan\lambda < 0. \quad (8.20)$$

For simplicity, in the remainder of this chapter, it is assumed that $c_x = 0$. Note that from (8.15), now we have $\mathrm{sgn}(N_0) = \mathrm{sgn}(R)$. In terms of systems' parameters, the equilibrium point is unstable whenever the following inequalities hold simultaneously:

$$\begin{aligned} \mu &> \tan\lambda, \\ R\Omega &> 0, \\ \tan\lambda\xi_+ m{r_m}^2 &> I. \end{aligned} \tag{8.21}$$

Expectedly, the instability conditions given by (8.21) are the same as the necessary conditions for the Painlevé's paradoxes discussed in Sect. 8.1. Limiting our study to the case of $\Omega > 0$ for simplicity, Fig. 8.7 shows that the phase plane of the system is divided by $N = 0$ and $\dot{\theta} = 0$ lines into four regions. In these regions, the system's equation has either no solution or two solutions when kinematic constraint instability is active (i.e., conditions of (8.21) are satisfied). Based on the discussions in Sect. 8.2, the following conclusions are drawn for the behavior of the lead screw model:

For any initial condition $[u(t_0), \dot{u}(t_0)]$, located in the hatched regions, the motion is seized instantaneously. This seizure is accompanied by an impulsive reaction force between contacting threads. The motion resumes from $[u(t_0), -\Omega]$ inside the unhatched region. This response also describes the trajectories that hit the $N(u, \dot{u}) = 0$ boundary. Inside the unhatched region, the trajectories follow $\Gamma_- \ddot{u} + c\dot{u} + ku = r_m(\xi_+ - \xi_-)R$ which is convergent to $u_0 = r_m(\xi_+ - \xi_-)(R/k)$.

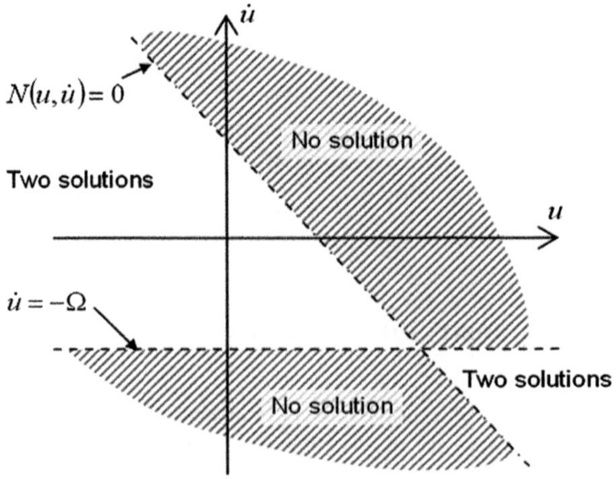

Fig. 8.7 Regions with no-solution and multiple solutions in the phase space of the system for $\mu > \tan\lambda$, $R > 0$, $\Omega > 0$, and $\Gamma_+ < 0$

8.6.1 Numerical Simulation Results

Solving $\Gamma_+ = 0$ for m, the critical translating mass is found to be $m = m_{cr} \approx 10.10\,\text{kg}$ for the system parameters given in Table 8.1. Figure 8.8 shows the evolution of the real and imaginary parts of the two eigenvalues given by (8.21) as the translating mass, m, is varied. For $m > m_{cr}$, the linearized system loses stability. Figure 8.9a, b show the phase plane plots of the 1-DOF model for $m = 10\,\text{kg}$ and $m = 11\,\text{kg}$, respectively. It can be seen that, by crossing the kinematic instability threshold, the origin becomes unstable.

To see what happens during the "sprag" phase, the same system parameters are used in the numerical simulation using the 2-DOF model of Sect. 5.5 with $T_0 = F_0 = 0$. In this example, very high contact stiffness and damping values are selected; $k_c = c_c = 10^8$. The projection of the trajectories for the lead screw DOF is shown in Fig. 8.10a which is almost indistinguishable from the 1-DOF system trajectories plotted in Fig. 8.9b. The impulse-like peaks in the contact force as the system goes through the "sprag" phase is shown in Fig. 8.10b. For the selected values of the contact stiffness and damping, this force peaks to about 320 kN.

As mentioned earlier, damping does not affect the stability of the 1-DOF model when the kinematic constraint instability mechanism is active. However, damping has a considerable effect on the behavior of the nonlinear system. Figure 8.11 shows phase plots of the 1-DOF model with three levels of lead screw support damping. For each of these three simulation results, the line $N(u, \dot{u}) = 0$ is also drawn. The onset of lead screw seizure is the point where the trajectory reaches this line. Note that from (8.17), the line $N(u, \dot{u}) = 0$ is given by

$$\dot{u} = -\frac{k}{c}u - \left(\frac{I - m{r_m}^2 \tan \lambda \xi_0}{m r_m \tan \lambda}\right)R.$$

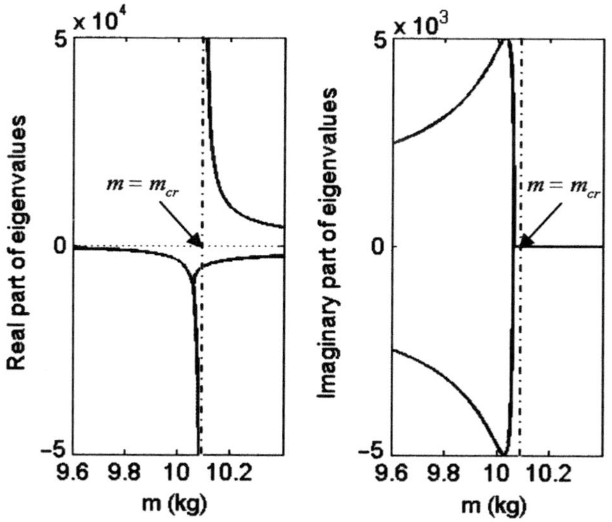

Fig. 8.8 Evolution of the eigenvalues as the translating mass, m, is varied

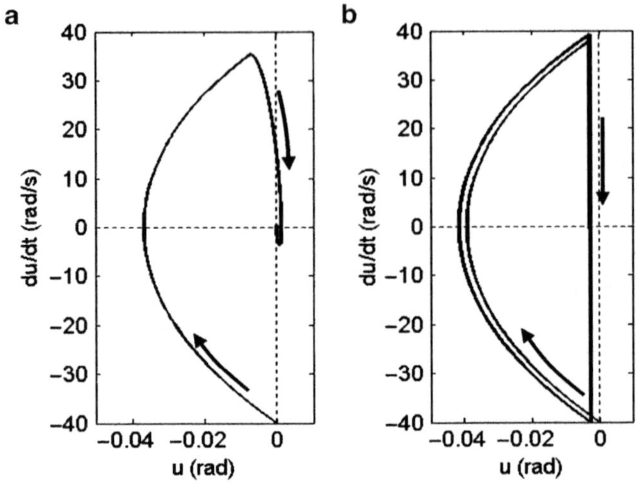

Fig. 8.9 System trajectories. (**a**) $m = 10 < m_{cr}$ (b) $m = 11 > m_{cr}$

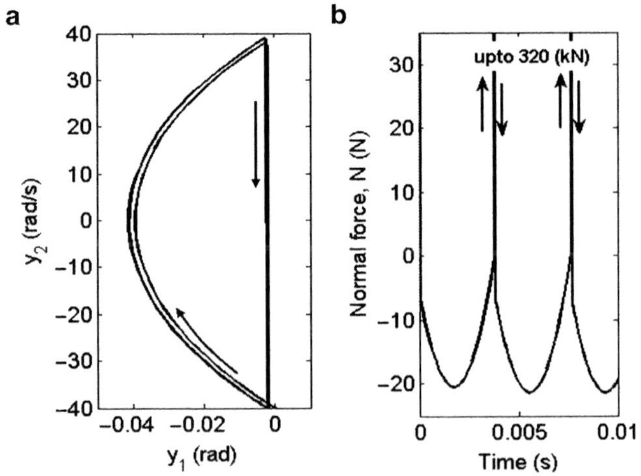

Fig. 8.10 Instability caused by kinematic constraint – 2-DOF model with very high contact stiffness and damping (**a**) phase-plane; (**b**) contact normal force

As damping is increased, the amplitude of vibrations is slightly reduced. However, as shown in Fig. 8.12, increasing damping increases the mean deflection of coupling element, which increases the mean thread normal force.

The results presented in Fig. 8.10 were obtained using a 2-DOF with very high contact stiffness and damping. As shown in Fig. 8.13, by decreasing the contact parameters (i.e., k_c and c_c) the trajectories become smoother and the deflection of the coupling element (i.e., torsional spring, k) may even become positive during the sprag phase.

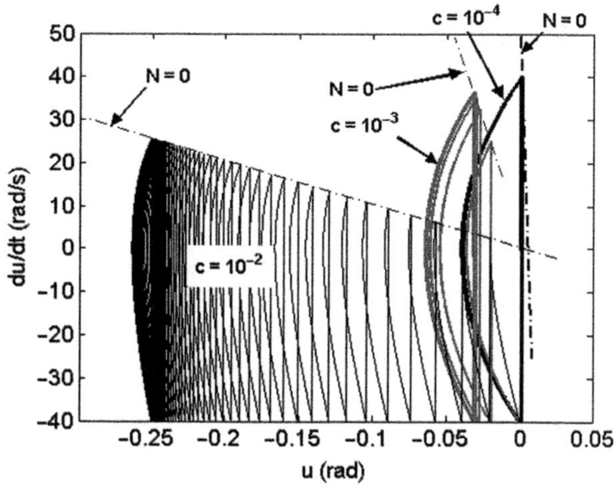

Fig. 8.11 Effect of damping

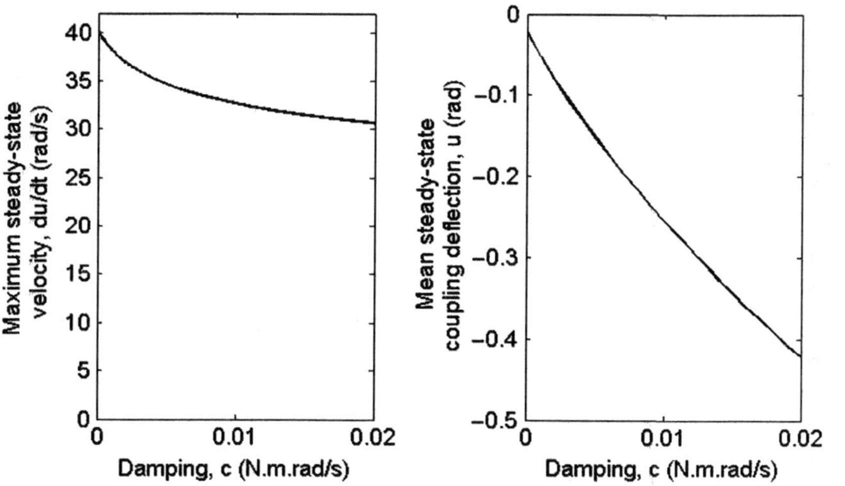

Fig. 8.12 Effects of damping on the steady-state vibration of the lead screw system under kinematic constraint instability

8.7 Region of Attraction of the Stable Equilibrium Point

The linear eigenvalue analysis of Sect. 8.6 showed that when $C_0, \Gamma_0 > 0$ the origin is stable. More specifically, when the kinematic constraint instability conditions given by (8.21) are *not* satisfied and $C_0 > 0$, the trivial equilibrium point of the system is asymptotically stable. However, there can be situations where the region of attraction of the stable equilibrium point is quite small, leading to instabilities even when conditions of (8.21) do not hold.

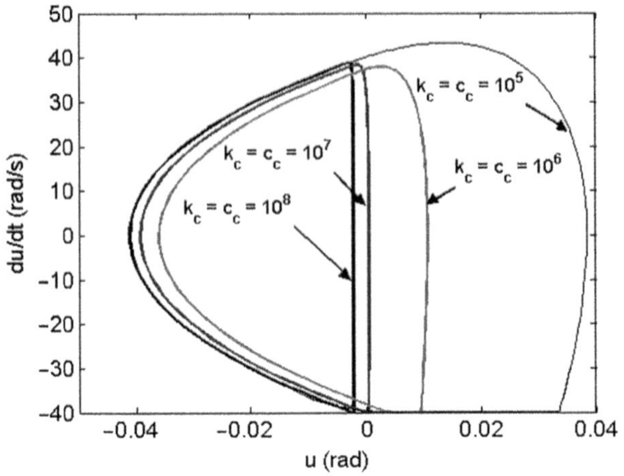

Fig. 8.13 Effect of contact parameters on the response of the system under kinematic constraint instability

Consider the case where $\mu > \tan\lambda$, $R < 0$, and $\Omega > 0$ (Note that in this case, we have; $\xi_{-} < 0 \rightarrow \Gamma_0 = \Gamma_{-} = I - \tan\lambda\xi_{-}mr_{\text{m}}{}^2 > 0$). It is obvious that only the first condition of (8.21) is satisfied and hence the steady-sliding equilibrium point is stable. Further, assume that $\Gamma_{+} = I - \tan\lambda\xi_{+}mr_{\text{m}}^2 < 0$ (this is the third condition of (8.21) if $R > 0$). Consider the initial conditions $(u(0), \dot{u}(0))$ such that $\Gamma_{-}R + mr_m \tan\lambda(ku(0) + c\dot{u}(0)) > 0$ $(i.e.\ N > 0)$ and $\dot{u}(0) \geq -\Omega(i.e.\ \dot{\theta} > 0)$. No motion is possible from this initial condition and the velocity goes to zero instantaneously $(i.e.\ \dot{u}(0^+) = -\Omega)$. Starting from this point, N is positive and the system dynamics is governed by the stable differential equation

$$\Gamma_{-}\ddot{u} + c\dot{u} + ku = 0. \tag{8.22}$$

The system's trajectory follows a path below the $\dot{u} = -\Omega$ line (i.e., reversed rotation of the lead screw) until it reaches the $N = 0$ line again and the lead screw rotation is once again seized. This cycle continues until the point $((-\Gamma_{-}R/kmr_m \tan\lambda + c\Omega/k), -\Omega)$ is reached, where the $N = 0$ line intersects the horizontal $\dot{u} = -\Omega$ line. Also note that initial motion from conditions where $\Gamma_{-}R + mr_{\text{m}} \tan\lambda(ku(0) + c\dot{u}(0)) < 0$ $(i.e.\ N < 0)$ and $\dot{u}(0) < -\Omega$ $(i.e.\ \dot{\theta} < 0)$ is also not possible and the system's trajectory transfers instantaneously to $(u(0), -\Omega)$ from which the motion is governed by (8.22) and continues towards origin with negative N and $\dot{u}(t) + \Omega > 0$.

Since (8.22) has an exponentially stable equilibrium point at the origin, all solutions that start from initial conditions, satisfying $\Gamma_{-}R + mr_{\text{m}} \tan\lambda(ku(0) + c\dot{u}(0)) \leq 0$ and $\dot{u}(0) \geq -\Omega$ and do not touch the $N = 0$ line, reach the origin (steady-sliding state) exponentially. If any of these trajectories reach the $N = 0$ line say at $t = t_1$, then the motion stops instantaneously and starts from the rest at $(u(t_1), -\Omega)$. This pattern continues and may result in a limit cycle at steady state.

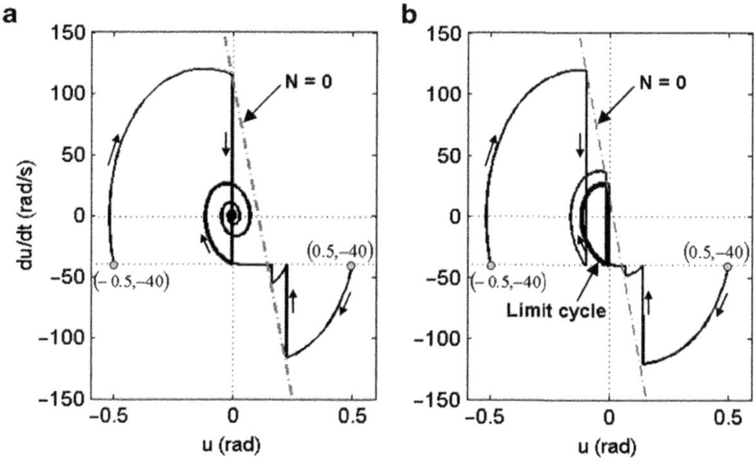

Fig. 8.14 System trajectories for $\Omega = 40$ (rad/s) (**a**) R $= -50(N)$; (**b**) R $= -10(N)$

Figure 8.14a shows two trajectories starting well away from the equilibrium point for $R = -50N$, $c = 10^{-3}$ N m s/rad, and $m = 10$ kg. Other system parameters are taken from Table 8.1. Although trajectories reach the $N = 0$ line, the origin is reached asymptotically. In Fig. 8.14b, the applied axial force is increased to $R = -10N$ while the other parameters are unchanged. In this case, the system trajectories are attracted to a limit cycle.

From the above discussions, one can conclude that, if $-R > 0$ is large enough such that every trajectory starting from $(u_0, -\Omega)$, where $u_0 \leq -(\Gamma\_R/kmr_{\mathrm{m}} \tan\lambda)+ (c\Omega/k)$ reaches the origin asymptotically, then the steady-sliding equilibrium point is globally stable. Otherwise, the region of attraction is only a subset of state space.

8.8 Kinematic Constraint Instability in Multi-DOF System Models

8.8.1 2-DOF Model of Sect. 5.6

The linearized equations of motion for this system were developed in Sect. 7.1.1. and the conditions for mode coupling instability were derived in the previous chapter. Here, we will only focus on the possibility of instability due to the kinematic constraint mechanism in the undamped system.

The natural frequencies of the undamped system are the roots of the following quadratic equation in ω^2:

$$\det\left(\mathbf{K} - \omega^2\mathbf{M}_1\right) = 0, \tag{8.23}$$

where **K** and $\mathbf{M}_1$ are given by (7.6) and (7.9), respectively. Expanding (8.23) yields

$$a_4\omega^4 + a_2\omega^2 + a_0 = 0,$$

where

$$\begin{aligned} a_4 &= I(m+m_1) - mm_1\xi_0 r_{\mathrm{m}}^2 \tan\lambda, \\ a_2 &= -k(m+m_1) - k_1\left(I - \xi_0 m r_{\mathrm{m}}^2 \tan\lambda\right), \\ a_0 &= kk_1, \end{aligned}$$

Since $a_0 > 0$, instability occurs whenever

$$a_4 < 0 \tag{8.24}$$

or

$$a_4 > 0 \wedge a_2 > 0 \tag{8.25}$$

or

$$a_2^2 - 4a_0a_4 < 0. \tag{8.26}$$

In terms of system parameters, the instability condition given by (8.24) can be written as

$$\tilde{\Gamma}_0 = I - \xi_0 \tan\lambda \tilde{m} r_{\mathrm{m}}^2 < 0, \tag{8.27}$$

where

$$\tilde{m} = \frac{mm_1}{m+m_1}. \tag{8.28}$$

Inequality (8.27) is the condition for the occurrence of Painlevé's paradoxes or the kinematic constraint instability.

Remark 7.2. Similar to the 1-DOF model in Sect. 8.4, the necessary condition for $\tilde{\Gamma}_0 < 0$ is $\xi_0 > 0$ which, in turn, requires $\mathrm{sgn}(R\Omega) = 1$ and $\mu > \tan\lambda$. □

Remark 7.3. Equation (8.27) takes the form of (8.20) with $\tilde{m}$ as the *equivalent* translating mass. □

Remark 7.4. Equation (8.27) is equivalent to requiring $A < 0$ where A is given by (5.38). □

The other two instability conditions (8.25) and (8.26), relate to the mode coupling instability mechanism and their analysis closely follows Sect. 7.2.1. Note that the situation $a_2 = 0$ satisfies (8.26), thus (8.25) does not define an instability boundary. The inequality (8.26) gives the necessary and sufficient for the mode coupling instability. Replacing the less-than sign with an equal sign for the instability boundary and after simplifications, one finds

$$b_1 k_1^2 + b_2 k k_1 + b_3 k^2 = 0, \tag{8.29}$$

where

$$\begin{aligned} b_1 &= \left(I - \xi_0 m r_{\mathrm{m}}^2 \tan\lambda\right)^2, \\ b_2 &= 2\left[m_1\left(\xi_0 m r_{\mathrm{m}}^2 \tan\lambda - I\right) - m\left(\xi_0 m r_{\mathrm{m}}^2 \tan\lambda + I\right)\right], \\ b_3 &= (m + m_1)^2. \end{aligned}$$

This equation is quadratic in k and k_1 and can be solved to find parametric relationships for the onset of the flutter instability. The conditions for the solutions to be real positive numbers are

$$b_2^2 - 4 b_1 b_3 \geq 0, \tag{8.30}$$

$$b_2 < 0. \tag{8.31}$$

In terms of system parameters, inequality (8.30) becomes

$$16\xi_0 m^2 r_{\mathrm{m}}^2 \tan\lambda (m + m_1)\left(I - \tilde{m}\xi_0 r_{\mathrm{m}}^2 \tan\lambda\right) \geq 0,$$

which yields

$$I - \xi_0 \tan\lambda \tilde{m} r_{\mathrm{m}}^2 \geq 0 \wedge \xi_0 > 0, \tag{8.32}$$

where $\tilde{m}$ is defined by (8.28). The second inequality given by (8.31), yields

$$\frac{m(m_1 - m)}{m + m_1} \xi_0 r_{\mathrm{m}}^2 \tan\lambda - I < 0,$$

which is satisfied whenever (8.32) is satisfied.

Lemma 8.2. *For the 2-DOF model of Sect. 5.6, the mode coupling and the kinematic constraint instability regions have no overlap in the parameter space.*

Proof. Compare (8.27) with (8.32). ■

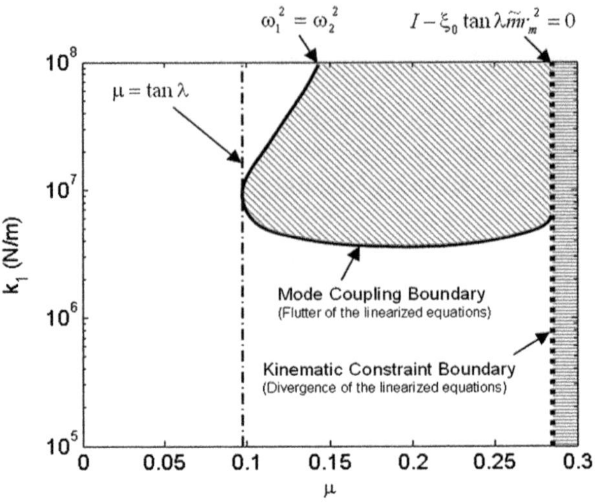

Fig. 8.15 Stability of the 2-DOF system as support stiffness and coefficient of friction are varied, where $m = 15$ kg, $m_1 = 11.6$ kg, and $R\Omega > 0$. *Slanted lines* hatched area: mode coupling instability region; *Horizontal lines* hatched area: primary kinematic constraint instability region

Figure 8.15 shows the instability regions of the undamped 2-DOF model in the $k_1 - \mu$ parameter space. Other system parameter values not given in the figure are selected according to Table 7.1. The two hatched regions correspond to the parameter range where mode coupling and kinematic constraint instability mechanisms are active.

As shown in this figure, for $m = 15$ and $m_1 = 11.6$, the condition (8.27) is satisfied for $\mu > \mu_{kc} \approx 0.285$. As a result, the vertical line $\mu \approx 0.285$ becomes the kinematic constraint instability boundary in the parameter space.

8.8.2 2-DOF Model of Sect. 5.8

The equations of motion of the model shown in Fig. 5.12 is given by (5.44) and (5.45). Let $\theta_i = \Omega t$ and

$$y_1 = \theta - \theta_i - \theta_0,$$
$$y_2 = \theta_2 - \theta_{20},$$

where

$$\theta_{20} = -r_m \frac{\xi_0}{k_2}(R - F_0 \mathrm{sgn}(\Omega)),$$

$$\theta_0 = -\frac{c\omega}{k} - r_m \frac{\xi_0}{k}[R - F_0 \mathrm{sgn}(\Omega)] - \frac{T_0 \mathrm{sgn}(\Omega)}{k}.$$

The equation of motion of this system in matrix form is derived as

$$\mathbf{M}\ddot{\mathbf{y}} + \mathbf{C}\dot{\mathbf{y}} + \mathbf{K}\mathbf{y} = \mathbf{F},$$

where $\mathbf{y} = [y_1 \quad y_2]^{\mathrm{T}}$ and

$$M = \begin{bmatrix} I - \xi m r_{\mathrm{m}}^2 \tan\lambda & \xi m r_{\mathrm{m}}^2 \tan\lambda \\ \xi m r_{\mathrm{m}}^2 \tan\lambda & I_2 - \xi m r_{\mathrm{m}}^2 \tan\lambda \end{bmatrix},$$

$$K = \begin{bmatrix} k & 0 \\ 0 & k_2 \end{bmatrix},$$

$$C = \begin{bmatrix} c & 0 \\ 0 & c_2 \end{bmatrix},$$

$$\mathbf{f} = \begin{bmatrix} r_{\mathrm{m}}R(\xi_0 - \xi) - r_{\mathrm{m}}F_0(\xi_0 \mathrm{sgn}(\Omega) - \xi \mathrm{sgn}(\dot{y}_1 - \dot{y}_2 + \Omega)) + T_0(\mathrm{sgn}(\Omega) - \mathrm{sgn}(\dot{y}_1 + \Omega)) \\ -r_{\mathrm{m}}R(\xi_0 - \xi) + r_{\mathrm{m}}F_0(\xi_0 \mathrm{sgn}(\Omega) - \xi \mathrm{sgn}(\dot{y}_1 - \dot{y}_2 + \Omega)) \end{bmatrix}.$$

The linearized equation of motion in matrix form in the neighbourhood of the steady-sliding equilibrium point (i.e. $\mathbf{y} = \dot{\mathbf{y}} = \mathbf{0}$) is

$$\mathbf{M}_0\ddot{\mathbf{y}} + \mathbf{C}\dot{\mathbf{y}} + \mathbf{K}\mathbf{y} = \mathbf{0},$$

where

$$\mathbf{M}_0 = \begin{bmatrix} I - \xi_0 m r_{\mathrm{m}}^2 \tan\lambda & \xi_0 m r_{\mathrm{m}}^2 \tan\lambda \\ \xi_0 m r_{\mathrm{m}}^2 \tan\lambda & I_2 - \xi_0 m r_{\mathrm{m}}^2 \tan\lambda \end{bmatrix}.$$

Because of the symmetry of the mass ($\mathbf{M}_0$), stiffness ($\mathbf{K}$), and damping ($\mathbf{C}$) matrices, the steady-sliding equilibrium point cannot loose stability by flutter. Divergence is also ruled out since $\det(\mathbf{K}) > 0$. As for the possibility of paradoxes, we set

$$\det(\mathbf{M}_0) = 0.$$

Consequently, kinematic constraint instability condition is found as

$$\frac{I_2 I}{I + I_2} - \xi_0 \tan\lambda m r_{\mathrm{m}}^2 < 0. \tag{8.33}$$

Remark 8.5. (8.33) is equivalent to requiring $A < 0$ where A is given by (5.51). □

Remark 8.6. Comparing (8.33) to (8.20) reveals that the addition of nut rotation DOF adversely affects the stability since the effective moment of inertia is reduced from I to $I_{\mathrm{eff}} = I_2 I/(I + I_2)$. □

8.9 Conclusions

In this chapter, the role of friction and the kinematic constraint equation (which defines the relative motion of lead screw and nut) in causing friction-induced vibrations in lead screw drives was investigated. Depending on the system parameters (including friction), the kinematic constraint may lead to instability in two distinct ways: Negative damping and the occurrence of Painlevé's paradoxes.

The conditions for the negative damping instability are given by (8.19). As for the Painlevé's paradoxes, the instability conditions are given by (8.21). It was found that self-locking condition as well as the application of the axial force in the direction of motion of the translating part are the two necessary conditions for instability. The true motion of the system in the region of paradoxes was established. Furthermore, it was shown that in both paradoxical situations (i.e., indeterminacy and inconsistency) tangential impact may occur, which results in discontinuous velocity changes in the rigid body model. The sprag-slip vibration caused by the kinematic constraint instability mechanism was studied numerically. Finally, two examples of the kinematic constraint instability in multi-DOF models of the lead screw drives were presented.

Chapter 9
An Experimental Case Study

In this chapter, [1] the friction-induced vibration of the lead screws incorporated in the horizontal drive mechanism of a type of powered seat adjuster is studied. The horizontal drive mechanism is constructed of two parallel lead screw slider systems. Torque is transmitted from a DC motor to the two lead screws through worm gearboxes. Flexible couplings connect the gearboxes to the motor and to the two lead screws. The nuts are stationary and are connected to the seat frame. The lead screw sliders together with the motor and gearboxes move with the seat as the lead screws advance in the nuts. In some cases, an extra force applied (by the passenger) in the direction of motion causes the system to generate audible noise, which is unacceptable to the car manufacturer.

In Sect. 9.1, some preliminary observations are made regarding the audible noise generated by the system. The mathematical model of a single slider mechanism is derived in Sect. 9.2. This model is the basis of a two-step parameter identification process that is described in Sects. 9.3 and 9.6. The developed identification method is capable of estimating various parameters quantifying friction, damping, and stiffness of the system. The test setup developed to perform detailed experiments on the lead screw drive is presented in Sect. 9.4. The experimental results corresponding to the first and second steps of the parameter identification are given in Sects. 9.5 and 9.7, respectively. These results clearly show that the mathematical model of the system together with the identified parameters can predict the behavior of the system. Parameter studies based on the mathematical model with identified parameters are given in Sect. 9.8. The conclusions are summarized in Sect. 9.9.

[1]Majority of the results presented in this chapter were previously published in 118.

O. Vahid-Araghi and F. Golnaraghi, *Friction-Induced Vibration in Lead Screw Drives*, DOI 10.1007/978-1-4419-1752-2_9,

9.1 Preliminary Observations

The first step toward finding the cause of the excessive audible noise in an operating mechanical system may be to analyze the noise signal and the external conditions under which such a noise is generated. In this section, some of the preliminary observations made on the powered seat adjuster are presented. First, a series of tests were performed on the complete seat adjuster. Figure 9.1 shows the test setup and the instrumentation used. A wall-mounted pneumatic cylinder was used to apply various levels of axial force to the seat adjuster's frame.

The instruments used, shown in Fig. 9.1c, were as follows:

- Force measurement: OMEGA®[2] pancake style LCHD 1,000 lb capacity
- Sound level (dBA) measurement: TES 1350A® Sound Level Meter[3]
- Audible noise (sound wave) recording: A general purpose PC microphone
- Seat displacement measurement: CELESCO®[4] position transducer SP1–12

Signals from the load cell, the position transducer, and the sound level meter were collected using a PC equipped with a Measurement Computing®[5] data acquisition card model PCI-DAS1602/16. A small Matlab®/Simulink® program was written to record signals received by the data acquisition card. A screenshot of the data acquisition program during one of the tests is shown in Fig. 9.1a. The sampling frequency was set to 1,000 Hz. The signal from the microphone was recorded by the Windows® standard sound recorder accessory software. The sound sampling frequency was 22,050 Hz.

A sample of these measurements is given in Fig. 9.2. As shown in Fig. 9.2c, an applied force of approximately 180 N caused the seat adjuster to generate audible noise. Figure 9.2d shows that seat was traveling at a velocity of approximately 7 mm/s. The sound level meter measurements in Fig. 9.2b shows an approximately 10 dB jump occurred in the noise level (compared to the background noise level) during a portion of the seat travel. During the same interval, the audible noise time-frequency plot in Fig. 9.2a clearly shows the sustained presence of a noise signal with a dominant frequency of approximately 160 Hz. The frequency content of the noise signal is also shown in Fig. 9.3 during an interval centered at $t = 8$ s.

The experiments performed on the complete seat adjuster were repeated for a single slider. Figure 9.4 shows the test setup for these tests. To simplify the test setup, the lead screw slider mechanism was installed upside-down compared to its configuration in the seat adjuster. In this setup, the DC motor rotates a single lead screw, which is horizontally fixed. As in the case of the complete seat experiments, a pneumatic cylinder applies the required axial force to the system. As shown in

[2]http://www.omega.com.

[3]http://www.tes.com.tw.

[4]http://www.celesco.com.

[5]http://www.measurementcomputing.com.

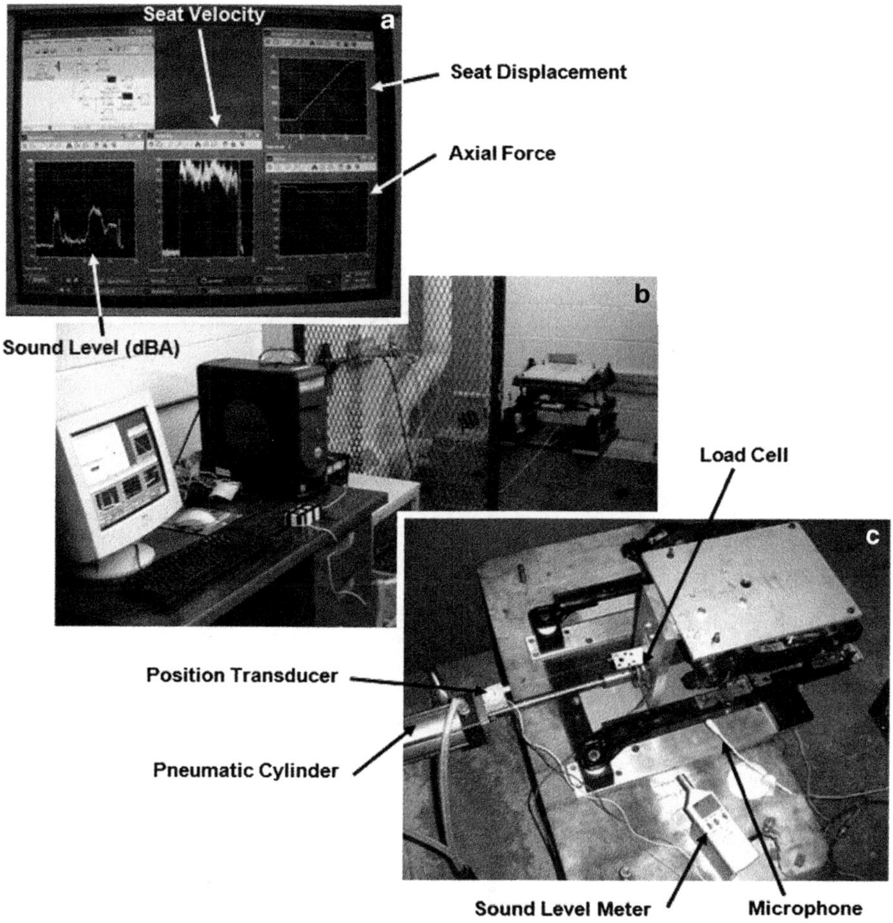

Fig. 9.1 Test setup for complete seat adjuster tests

Fig. 9.4, force is applied directly to the nut parallel to the lead screw axis. Instrumentations used were those used in the previous test setup.

Sample measurement results are presented in Fig. 9.5. In this test, a horizontal force of about 200 N (Fig. 9.5c) was needed to induce the noise at a traveling velocity of approximately 20 mm/s (Fig. 9.5d). The audible noise continued for about 4 s with a dominant frequency of about 150 Hz (Fig. 9.5a) accompanied by an almost 20 dB increase in the noise level (Fig. 9.5b). The frequency spectrum of the recorded noise at $t = 3$ s is plotted in Fig. 9.6. This plot clearly shows the dominant signal frequency of 150 Hz.

Consistency in the frequency of the generated audible noise between complete seat tests and single slider tests, points toward the lead screw vibrations as the source of the noise. In addition, a rough estimate of the natural frequency of

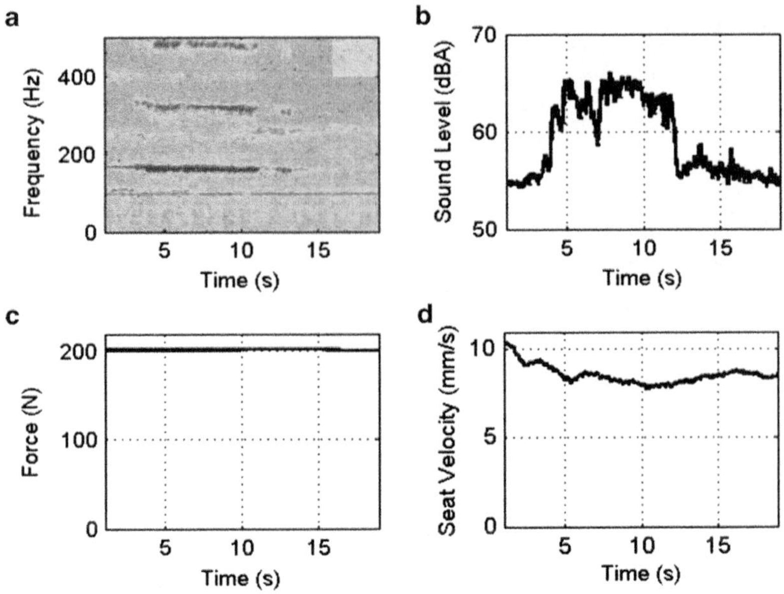

Fig. 9.2 Sample test results from complete seat adjuster tests

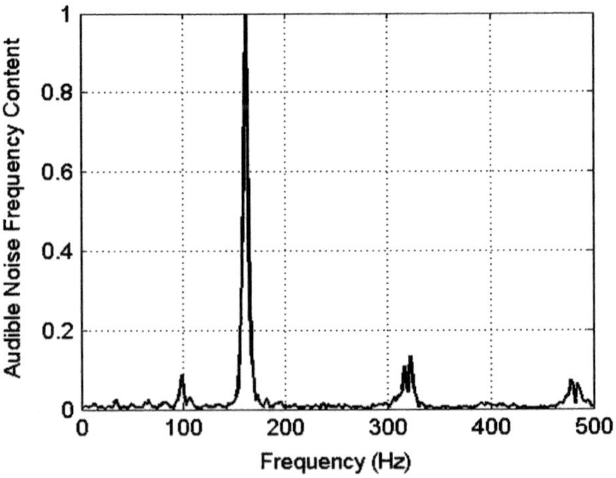

Fig. 9.3 Audible noise frequency content at 8 s for the test results shown in Fig. 9.2a. Peak amplitude at 162 Hz

the lead screw drive based on the model of Sect. 2.3 and known (i.e., from the manufacturer's data) and typical system parameters is found to be consistent with the above findings. The linear undamped natural frequency of lead screw is found from (5.18) to be

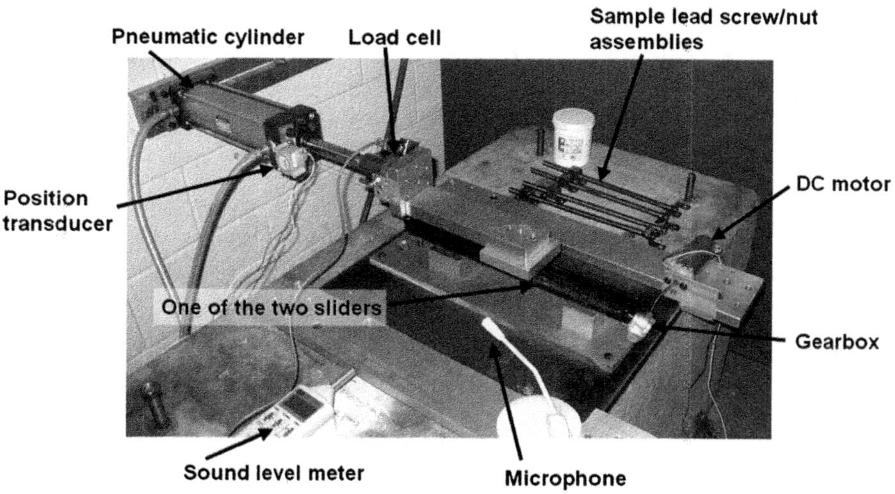

Fig. 9.4 Single-track test setup

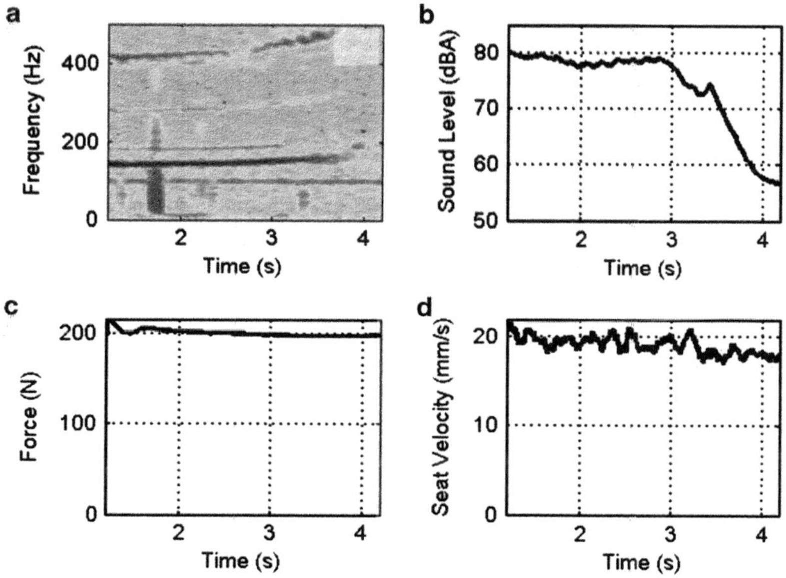

Fig. 9.5 Sample test results from single-track tests

$$f = \frac{1}{2\pi}\sqrt{\frac{k}{I - \tan\lambda\xi_0 m r_{\mathrm{m}}{}^2}}.$$

Let $I = 3.1 \times 10^{-6}\,\mathrm{kg\,m^2}$, $m = 4\,\mathrm{kg}$, $r_{\mathrm{m}} = 5.2 \times 10^{-3}\,\mathrm{m}$, and, $k = 3.4\,\mathrm{N\,m/rad}$. Also setting $\lambda = 0.1\,\mathrm{rad}$ and $\mu_0 = 0.2$ gives $\xi_0 \approx 0.1$, the natural frequency of

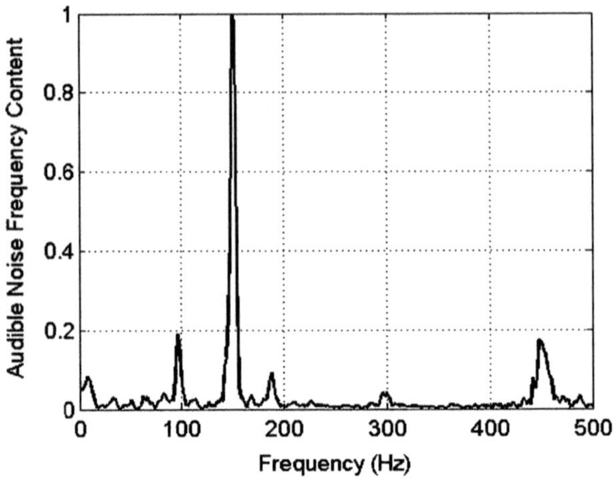

Fig. 9.6 Audible noise frequency content at 3 s for the test results shown in Fig. 9.5a. Peak amplitude at 150 Hz

vibrations from the above formula is estimated as $f \approx 205$ Hz. This value is higher than the dominant frequency of the measured audible noise (see Figs. 9.3 and 9.6). The difference may be explained by the fact that the flexible coupling with $k = 3.4\,\mathrm{Nm/rad}$ connects the lead screw to a gearbox with plastic worm and worm gear. The flexibility of gearbox thus reduces the overall effective torsional stiffness and consequently may account for the lower actual natural frequency of lead screw vibrations. Detail modeling of the gearbox, however, is not pursued further here. The effective torsional stiffness of the system is estimated experimentally as explained below.

In the remainder of this chapter, the focus is on a single slider mechanism. Based on the setup in Fig. 9.4, a new test setup is developed for the system parameter identification. The mathematical model of this setup is described next.

9.2 Mathematical Modeling

Figure 9.7 shows a schematic view of the parameter identification test setup designed to perform focused experiments on a single slider system. The mathematical model of this system is presented in this section. This model forms the basis of the parameter identification method to be described in Sects. 9.3 and 9.6 below. As shown in Fig. 9.7, two rotary encoders are used to measure the angular displacements of the lead screw, θ, and the motor, θ_{M}. A load cell is used to measure the force exerted by the pneumatic cylinder, R. The input current to the DC motor are also measured. The input torque to the system, T_{M}, is then calculated from this quantity and the known torque constant of the DC motor.

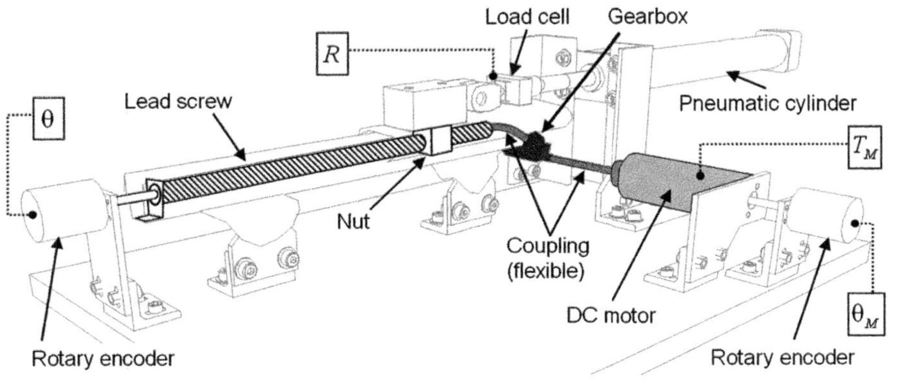

Fig. 9.7 Parameter identification test setup

Figure 9.8 shows the 3-DOF model of the parameter identification test setup. Similar to the system in Fig. 9.7, the model consists of three parts, i.e., DC motor, gearbox, and lead screw mechanism. These parts are connected to each other via couplings with torsional compliance.

For the motor, Newton's second law gives

$$I_{\mathrm{M}}\ddot{\theta}_{\mathrm{M}} = T_{\mathrm{M}} - T_{\mathrm{fM}}\mathrm{sgn}\dot{\theta}_{\mathrm{M}} - c_{\mathrm{M}}\dot{\theta}_{\mathrm{M}} + k_1(\theta_{\mathrm{W}} - \theta_{\mathrm{M}}), \tag{9.1}$$

where I_{M} is the moment of inertia of the rotor, T_{fM} and c_{M} are the internal friction and damping of the motor, respectively. Also, k_1 is the torsional stiffness of the coupling connecting the motor to the gearbox and θ_{W} designates the angular displacement of the worm.

For the worm and worm gear, Newton's second law gives

$$\begin{aligned} I_{\mathrm{W}}\,\ddot{\theta}_{\mathrm{W}} = {} & k_1(\theta_{\mathrm{M}} - \theta_{\mathrm{W}}) - c_{\mathrm{W}}\dot{\theta}_{\mathrm{W}} - \frac{d_{\mathrm{W}}}{2}\left(W\sin\lambda_{\mathrm{W}} + \mu_{\mathrm{WG}}|W|\mathrm{sgn}\dot{\theta}_{\mathrm{W}}\cos\lambda_{\mathrm{W}}\right) \\ & - T_{\mathrm{fW}}\mathrm{sgn}\dot{\theta}_{\mathrm{W}}, \end{aligned} \tag{9.2}$$

$$\begin{aligned} I_{\mathrm{G}}\,\ddot{\theta}_{\mathrm{G}} = {} & k(\theta - \theta_{\mathrm{G}}) - c_{\mathrm{G}}\dot{\theta}_{\mathrm{G}} + \frac{d_{\mathrm{G}}}{2}\left(W\cos\lambda_{\mathrm{W}} - \mu_{\mathrm{WG}}|W|\mathrm{sgn}\dot{\theta}_{\mathrm{G}}\sin\lambda_{\mathrm{W}}\right) \\ & - T_{\mathrm{fG}}\mathrm{sgn}\dot{\theta}_{\mathrm{G}}, \end{aligned} \tag{9.3}$$

where I_{W} and I_{G} are the moment of inertia of the worm and worm-gear, respectively. θ_G is the angular displacement of the gear. d_{W} and d_{G} are the pitch diameters of worm and worm gear, respectively. W is the normal component of the contact force between meshing worm threads and gear teeth. T_{fW} and T_{fG} are the internal friction torques of the worm and the worm gear, respectively. λ_{W} is the pitch angle of the worm and μ_{WG} is the coefficient of friction of the meshing worm and worm gear.

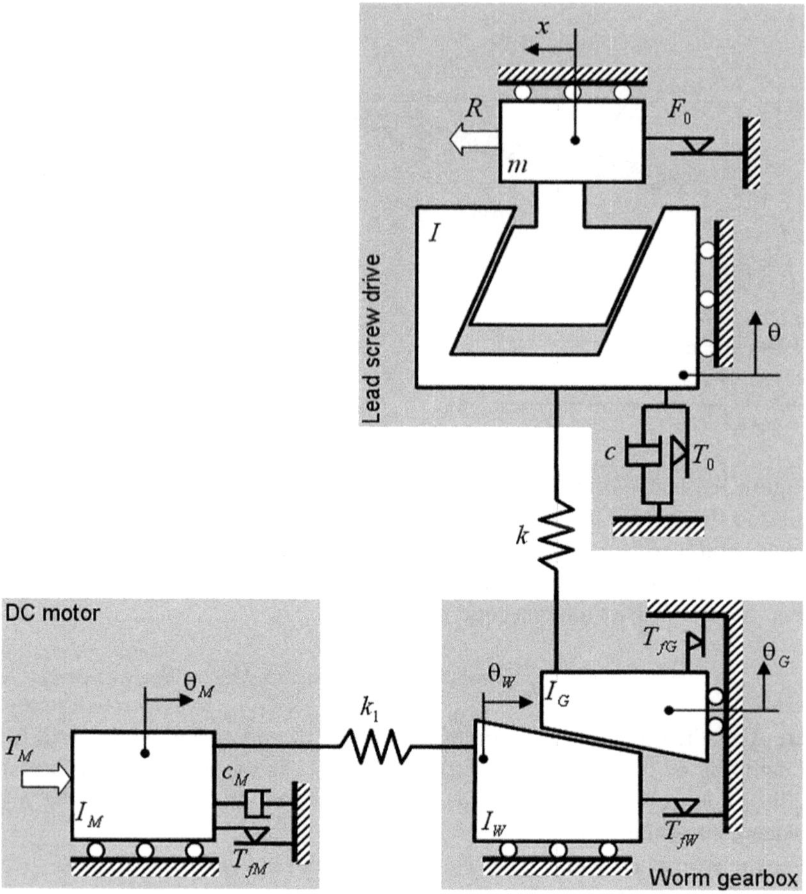

Fig. 9.8 3-DOF model of the lead screw test setup

Eliminating W between (9.2) and (9.3) gives

$$\begin{aligned}(I_W + \alpha_W\xi_W I_G)\ddot{\theta}_G + (c_W + \alpha_W\xi_W c_G)\dot{\theta}_G + (k_1 + \alpha_W\xi_W k)\theta_G \\ -k_1\alpha_W\theta_M - \alpha_W\xi_W k\theta + (\alpha_W T_{fW} + \alpha_W\xi_W T_{fG})\mathrm{sgn}\dot{\theta}_G = 0,\end{aligned} \tag{9.4}$$

where

$$\alpha_W = \frac{d_W}{d_G}\tan\lambda_W \tag{9.5}$$

is the gearbox ratio[6] (i.e., $\theta_G = \alpha_W\theta_W$) and

[6] For the worm gearbox considered here, $\alpha_W = n_W/n_G$ where n_G is the number of gear teeth and n_W is the number of worm starts.

$$\xi_{\mathrm{W}} = \frac{d_{\mathrm{W}}(\tan\lambda_{\mathrm{W}} + \mu_{\mathrm{WG}}\mathrm{sgn}(\dot{\theta}_{\mathrm{W}}W))}{d_{\mathrm{G}}(1 - \mu_{\mathrm{WG}}\mathrm{sgn}(\dot{\theta}_{\mathrm{G}}W)\tan\lambda_{\mathrm{W}})}.$$

For the lead screw and nut, the equation of motion is given by (5.18) which is repeated here for ease of reference.

$$(I - \tan\lambda\xi m r_{\mathrm{m}}{}^2)\ddot{\theta} + k\theta + c\dot{\theta} = k\theta_{\mathrm{G}} - r_{\mathrm{m}}\xi(R - F_0\mathrm{sgn}(\dot{\theta})) - T_0\mathrm{sgn}(\dot{\theta}), \quad (9.6)$$

where

$$\xi = \frac{\mu\mathrm{sgn}(N\dot{\theta}) - \tan\lambda}{1 + \mu\mathrm{sgn}(N\dot{\theta})\tan\lambda}. \quad (9.7)$$

In (9.7), N is the normal contact force between lead screw and nut threads and is given by (5.19). Also, the coefficient of friction is considered here to be a function of the relative sliding velocity (which is expressed more conveniently as a function lead screw angular velocity) and is given by (5.3) and is repeated here;

$$\mu = \mu_1 + \mu_2 \mathrm{e}^{-r_0|\dot{\theta}|} + \mu_3|\dot{\theta}|. \quad (9.8)$$

The equations of motion for the 3-DOF model in Fig. 9.8 are given by (9.1), (9.4), and (9.6).

Remark 9.1. In the actual lead screw system, lubrication, surface conditions, and load distribution on the lead screw and nut threads change as the nut translates along the lead screw. However, in the mathematical modeling and the subsequent identification approach, all parameters are taken as their averaged values over a small travel distance of the nut along the lead screw. □

9.3 Parameter Identification Step 1: Friction and Damping

In the first step of the parameter identification approach, the steady-state pure-slip conditions are considered (i.e., no rotational vibrations and constant lead screw angular velocity). The vibration-free operation of the system may be achieved through feedback control. Based on the mathematical model of the system developed in the preceding section, steady-state relationships are derived and by relating the measurable system inputs and states to the internal friction and damping parameters, these parameters are estimated. Table 9.1 lists the measured (or calculated) quantities and the main parameters to be identified in this step.

Table 9.1 Parameters of interest in step 1 of identification

Measured/calculated			To be estimated	
Parameter	Description		Parameter	Description
			μ_1	Parameter related to Coulomb friction
ω_M	Motor angular velocity		μ_2	Parameter related to Stribeck friction
		⇨		
T_M	Motor torque		μ_3	Parameter related to viscous friction
R	Axial force		r_0	Parameter related to Stribeck friction
			c	Damping coefficient of the lead screw support

In the second step of the parameter identification approach – described in Sect. 9.6 – the vibratory response of the system is used to fine-tune the friction and damping parameters and to identify overall torsional stiffness of the system by matching the response of the mathematical model and the measurements. The measured/calculated quantities in Table 9.1 are assumed available as their averaged values over the considered travel stroke of the nut.

The steady-state relationships are derived from (9.1), (9.4), and (9.6) by setting all accelerations to 0 and assuming positive constant angular velocities. One finds

$$T_M - T_{fM} + k_1\left(\alpha_W^{-1}\theta_G - \theta_M\right) - c_M\omega_M = 0, \tag{9.9}$$

$$\begin{aligned}(c_W + \alpha_W\xi_{W0}c_G)\omega_G + (k_1 + \alpha_W\xi_{W0}k)\theta_G - k_1\alpha_W\theta_M \\ -\alpha_W\xi_{W0}k\theta + (\alpha_W T_{fW} + \alpha_W\xi_{W0}T_{fG}) = 0,\end{aligned} \tag{9.10}$$

$$k(\theta_G - \theta) - c\omega - r_m\xi_0(R - F_0) - T_0 = 0, \tag{9.11}$$

where $\omega = \omega_G = \alpha_W\omega_M$ are the constant angular velocities and

$$\xi_{W0} = \frac{d_W(\tan\lambda_W + \mu_{WG})}{d_G(1 - \mu_{WG}\tan\lambda_W)} \tag{9.12}$$

and

$$\xi_0 = \frac{\mu\,\mathrm{sgn}(N) - \tan\lambda}{1 + \mu\,\mathrm{sgn}(N)\tan\lambda}. \tag{9.13}$$

The current study is limited to cases where the axial load is applied in the direction of motion, thus only cases where $R > 0$ are considered. Furthermore, F_0 is assumed to be negligible compared to R and as a result, N is assumed to be positive at steady-state conditions. Thus (9.13) is simplified to

$$\xi_0 = \frac{\mu - \tan\lambda}{1 + \mu\tan\lambda} \approx \mu - \tan\lambda, \tag{9.14}$$

where the approximation is obtained by assuming $\mu \tan\lambda \ll 1$. Combining (9.9)–(9.11) yields

$$T_{\mathrm{M}} = C\omega_{\mathrm{M}} + T_{\mathrm{f}} + r_{\mathrm{m}}\xi_{\mathrm{WO}}\xi_{\mathrm{O}}(R - F_{\mathrm{O}}), \tag{9.15}$$

where

$$C = C_0 + \alpha_W \xi_{WO} c, \tag{9.16}$$

$$T_f = T_{f0} + \xi_{WO} T_{\mathrm{O}} \tag{9.17}$$

and

$$C_0 = c_W + \alpha_W \xi_{WO} c_G + c_M, \tag{9.18}$$

$$T_{\mathrm{f0}} = T_{\mathrm{fM}} + T_{\mathrm{fW}} + \xi_{\mathrm{WO}} T_{\mathrm{fG}}. \tag{9.19}$$

To separate the *force* effects from the *velocity* effects, (9.15) is rearranged as

$$T_{\mathrm{M}} = \beta_0(\omega_{\mathrm{M}}) + (R - F_{\mathrm{O}})\,\beta_1(\omega_{\mathrm{M}}), \tag{9.20}$$

where

$$\beta_0(\omega_{\mathrm{M}}) = T_{\mathrm{f}} + C\omega_{\mathrm{M}}, \tag{9.21}$$

$$\beta_1(\omega_{\mathrm{M}}) = r_{\mathrm{m}}\xi_{\mathrm{WO}}\xi_{\mathrm{O}}(\omega_{\mathrm{M}}). \tag{9.22}$$

For each motor speed setting, $\omega_{\mathrm{M}}^{(i)}$, the straight line described by (9.20) can be fitted to the experimental data points $\left\langle R^{(j)}, T_{\mathrm{M}}^{(i,j)} \right\rangle$ to obtain $\beta_0^{(i)}$ and $\beta_1^{(i)}$ as functions of the motor angular velocity.

Based on (9.21), another straight line can be fitted to $\left\langle \omega_{\mathrm{M}}^{(i)}, \beta_0^{(i)} \right\rangle$ data points to obtain T_{f} and C. Expanding the second velocity-dependent coefficient, $\beta_1(\omega_{\mathrm{M}})$, using (9.14), gives

$$\beta_1(\omega) = r_{\mathrm{m}}\xi_{\mathrm{WO}}(\mu(\omega) - \tan\lambda). \tag{9.23}$$

At steady-state velocity of $\omega = \alpha_{\mathrm{W}}\omega_{\mathrm{M}}$, the coefficient of friction defined by (9.8) becomes

$$\mu = \mu_1 + \mu_2 e^{-r_0 \alpha_{\mathrm{W}} \omega_{\mathrm{M}}} + \mu_3 \alpha_{\mathrm{W}} \omega_{\mathrm{M}}. \tag{9.24}$$

Substituting (9.24) into (9.23) and rearranging gives

$$\beta_1(\omega_{\mathrm{M}}) = \begin{bmatrix} 1 & e^{-r_0 \alpha_{\mathrm{W}} \omega_{\mathrm{M}}} & \omega_{\mathrm{M}} \end{bmatrix} \begin{bmatrix} \gamma_0 \\ \gamma_1 \\ \gamma_2 \end{bmatrix}, \tag{9.25}$$

where

$$\gamma_0 = r_\mathrm{m}\xi_\mathrm{W0}(\mu_1 - \tan\lambda), \tag{9.26}$$

$$\gamma_1 = r_\mathrm{m}\xi_\mathrm{W0}\mu_2, \tag{9.27}$$

$$\gamma_2 = r_\mathrm{m}\xi_\mathrm{W0}\alpha_\mathrm{W}\mu_3. \tag{9.28}$$

Now the curve described by (9.25) can be fitted to the previously obtained data points (i.e., $\left\langle \omega_\mathrm{M}^{(i)}, \beta_1^{(i)} \right\rangle$) to estimate the three new parameters: $\hat{\gamma}_0$, $\hat{\gamma}_1$, and $\hat{\gamma}_2$. Using the least squares technique, one finds

$$\hat{\mathbf{\Gamma}} = \left(\mathbf{A}^\mathrm{T}\mathbf{A}\right)^{-1}\mathbf{A}^\mathrm{T}\mathbf{B}, \tag{9.29}$$

where

$$\hat{\mathbf{\Gamma}} = [\hat{\gamma}_0 \quad \hat{\gamma}_1 \quad \hat{\gamma}_2]^\mathrm{T}$$

and

$$\mathbf{A} = \begin{bmatrix} 1 & e^{-r_0\alpha_\mathrm{W}\omega_\mathrm{M}^{(1)}} & \omega_\mathrm{M}^{(1)} \\ 1 & e^{-r_0\alpha_\mathrm{W}\omega_\mathrm{M}^{(2)}} & \omega_\mathrm{M}^{(2)} \\ \vdots & \vdots & \vdots \\ 1 & e^{-r_0\alpha_\mathrm{W}\omega_\mathrm{M}^{(n)}} & \omega_\mathrm{M}^{(n)} \end{bmatrix}, \tag{9.30}$$

$$\mathbf{B} = \left[\beta_1^{(1)} \quad \beta_1^{(2)} \quad \cdots \quad \beta_1^{(n)}\right]^\mathrm{T},$$

where n is the total number of data points available. Note that $\mathbf{A}$ given by (9.30) is dependent on r_0, which is one of the unknown parameters describing the Stribeck effect in the assumed model of the velocity-dependent coefficient of friction. To solve this problem, a simple optimization routine is used to find the best value for r_0 such that the curve fitting error of (9.29) is minimized. Define

$$\begin{bmatrix} e_{(1)}(r_0) \\ e_{(2)}(r_0) \\ \vdots \\ e_{(n)}(r_0) \end{bmatrix} = \mathbf{B} - \mathbf{A}(r_0)\begin{bmatrix} \hat{\gamma}_0(r_0) \\ \hat{\gamma}_1(r_0) \\ \hat{\gamma}_2(r_0) \end{bmatrix},$$

where dependence on r_0 is made explicit. The optimized value of r_0 given by $\hat{r}_0$ is now found as

$$\hat{r}_0 = \underset{r_{0\,\min} \le r_0 \le r_{0\,\max}}{\arg\min} \sum_{i=1}^{n} e_{(i)}^2(r_0).$$

This task may be performed using a number of numerical optimization techniques. Here $\sum_{i=1}^{n} e_{(i)}^2(r_0)$ is simply computed and plotted over a range of appropriate values ($r_0 > 0$) and the minimum is found graphically.

9.4 Parameter Identification Test Setup

The test setup used in the friction identification experiments is shown in Fig. 9.9. Similar to the test setup shown in Fig. 9.4, only one of the two sliders is included in the setup. The working parts of the test setup are taken from an actual seat adjuster. Two encoders are used to measure the angular displacement of the lead screw and the motor. A load cell is used to measure the force exerted by the pneumatic cylinder. The input voltage and current to the DC motor are also measured. With the help of a controller regulating the current input to the DC motor [119, 120], the slider is set to move at near constant preset velocities in the applicable range. The angular velocity of the motor is calculated by numerical differentiation of its measured angular displacement. The motor torque is calculated from the measured input current and the known motor's torque constant. See Table 9.2 for a list of instruments and components of this test setup.

In the closed-loop tests, the DC motor is driven through a servo amplifier operating in the "current mode." In this mode, the current output of the amplifier is proportional to the input voltage control signal. Consequently, the motor torque is proportional to the control signal. The amplifier gain and the DC motor torque constant are 1.0 (A/V) and 0.0266 (N m/A), respectively. The control signal for the closed-loop tests is generated by a dSpace®[7] controller, which is programmed in Matlab. The pneumatic cylinder is activated by a solenoid valve, which is also commanded by the controller. Two identical analog rotary encoders (sinusoidal signal, 1 V_{pp}) are used to measure the angular displacement of the lead screw and the DC motor. These encoders have a resolution of 3,600 counts per revolution, which is interpolated up to 4,000 times by the dSpace controller and recorded. Other measured signals in these tests are the load cell signal (applied axial force) and the motor current, which are also acquired by the dSpace system.

Figure 9.10 shows a sample of measured angular displacement and calculated angular velocity (by numerical differentiation) of the lead screw.

The measurement data corresponding to the accelerating (start of motion) and decelerating (end of motion) portions of each test is discarded and the resulting near

[7]http://www.dspaceinc.com.

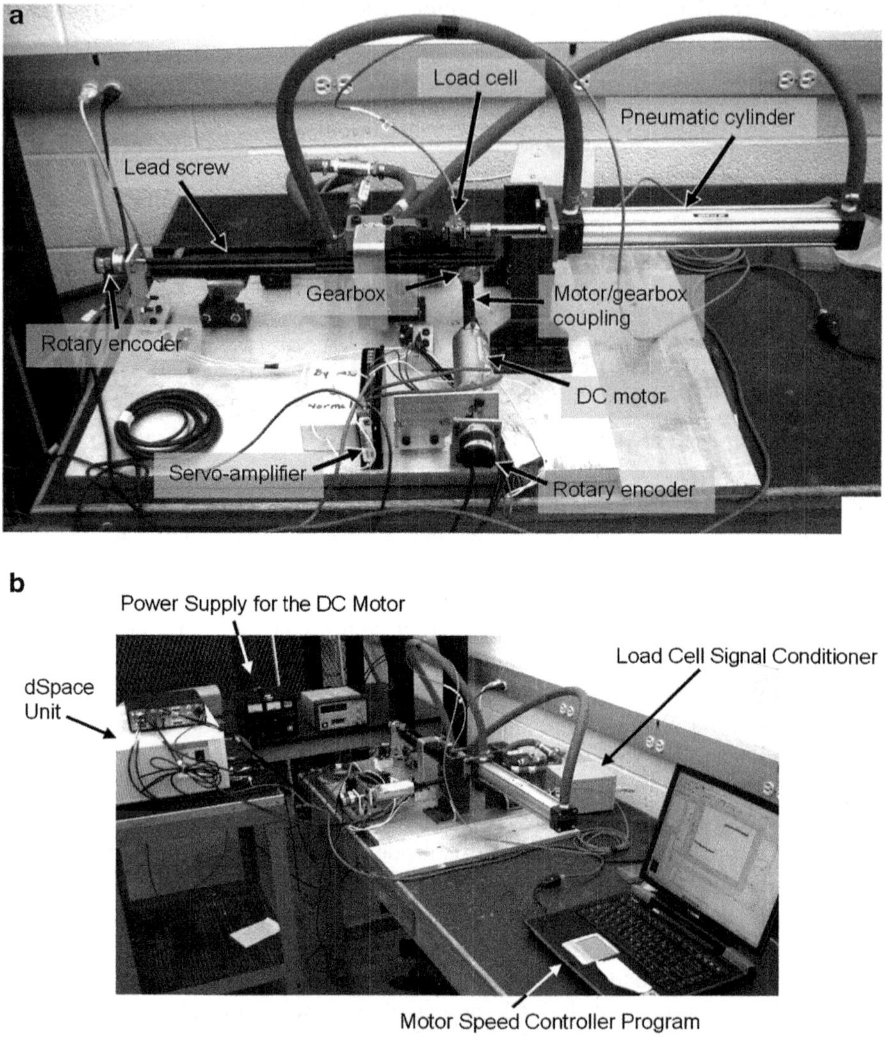

Fig. 9.9 Parameter identification test setup

steady-state measurements is averaged and recorded as one data point. See Fig. 9.11 for a sample of the near steady-state measurement results.

9.5 Experimental Results: Step 1

In this section, the parameter identification approach described in Sect. 9.3 is applied to the measurements performed using the test setup described in the previous section.

Table 9.2 Partial list of components of the lead screw test setup

No.	Item	Specification	Model	Manufacturer
1	Rotary encoders	3,600 lines per revolution, sinusoidal incremental signal (1 Vpp)	ERN 1080	Heidenhain http://www.heidenhain.com
2	Load cell	200 lbf Mini Universal Link Load Cell	LC703–200	Omega http://www.omega.com
3	Load cell signal conditioner	Strain gage amplifier	DMD-465	Omega http://www.omega.com
4	Motor servo amplifier	Pulse width modulation amplifier	12A8M	Advance Motion Control http://www.a-m-c.com
5	Power supply	DC-regulated power supply	–	BK Precision http://www.bkprecision.com
6	Solenoid valve	4 way, 2 solenoids valve with center exhaust	MVSC 300 4E2R	Mindman Pneumatics http://www.mindman.com.tw
7	Pneumatic cylinder	Double acting with $11^{3/4\prime\prime}$ stroke	MCQNF 11–1.5–1175	Mindman Pneumatics http://www.mindman.com.tw

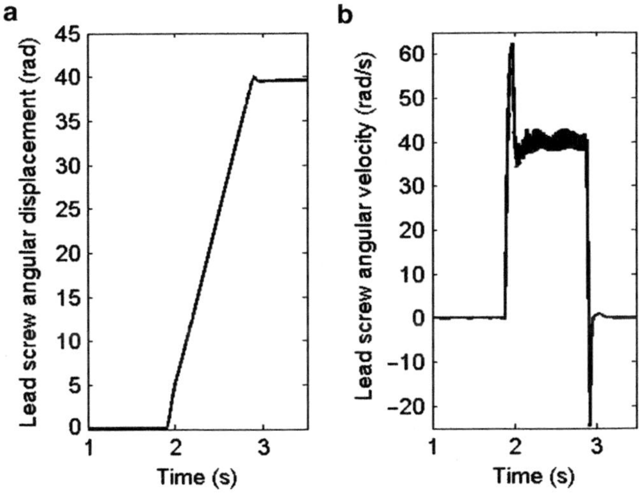

Fig. 9.10 Sample test results. (**a**) Lead screw angular displacement, (**b**) lead screw angular velocity

Before exploring the friction torque produced at the contact between lead screw and nut threads, preliminary measurements are required to estimate and isolate internal damping (9.18) and friction (9.19) of the DC motor and the gearbox. In

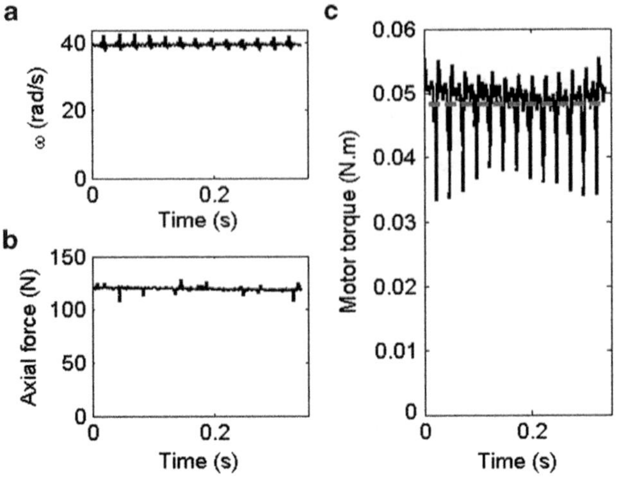

Fig. 9.11 Near steady-state portion of a sample test results. (**a**) Lead screw angular velocity, (**b**) axial load, and (**c**) motor torque

Sect. 9.5.1, the results of these calculations are presented and then, in Sect. 9.5.2, the lead screw friction and damping identification results are given.

9.5.1 DC Motor and Gearbox

In a series of preliminary tests, DC motor and gearbox were disconnected from the lead screw, and the input current of the DC motor was measured at different levels of preset constant angular velocities. Figure 9.12 shows the results of these tests.

By fitting a straight line to these data points using the least squares technique, the overall damping, C_0, and residual friction torque, T_{f0}, were estimated. These results together with other known system parameters are listed in Table 9.3.

9.5.2 Identification Results

Figure 9.13 shows data points collected from some of the measurements performed. In this figure, motor torque (measured from motor input current) is plotted against measured force and measured speed. Fluctuation in the supply air pressure to the cylinder, together with the speed-dependent internal friction of the piston rod, caused variations in the applied force from one experiment to the next.

As described in the previous section, a straight line is fitted to the data points at each velocity setting, which gives variation of motor torque vs. applied axial force

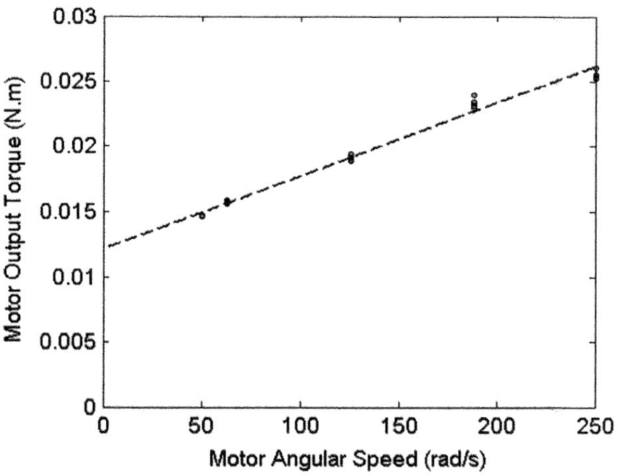

Fig. 9.12 Resistive torque of the motor and the gearbox. *Dots*: measurements, *dashed line*: fitted line to the data points

Table 9.3 Known or assumed system parameter values

Parameter	Value
Lead screw pitch diameter, d_m	10.366 mm
Lead screw lead angle, λ	5.57°
Mass of translating parts, m	3.8 kg
Average resistance of the slider, F_0	< 2 N
Assumed contact stiffness – lead screw and nut, k_c	10^8 N/m
Assumed contact damping – lead screw and nut, c_c	10^6 kg/s
Lead screw moment of inertia, I	3.12×10^{-6}kg m^2
Worm pitch diameter, d_W	9.442 mm
Worm gear pitch diameter, d_G	20.04 mm
Worm lead angle, λ_W	18.53°
Gearbox ratio, α_W	3/19
Nominal torsional stiffness of the coupling, k	1.12 N m/rad
Assumed coefficient of friction of gearbox mesh, μ_{WG}	0.2
Overall DC motor and the gearbox internal damping, C_0	0.0121 N m
Overall DC motor and the gearbox internal friction, T_{f0}	5.61×10^{-5}N m rad/s

according to (9.20) for each of the available velocity set points. Figure 9.14 shows a few samples of these curve fittings[8]. The curve fitting results according to (9.21) and (9.25) are shown in Figs. 9.15 and 9.16, respectively. The estimated parameters are listed in Table 9.4.

[8]In these calculations, the effect of F_0 was neglected since preliminary observations showed that the slider friction force is consistently less than 2 N, which is less than 2% of the applied force, R.

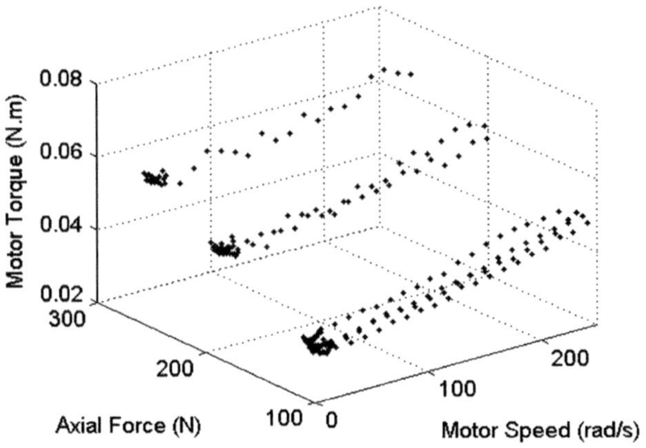

Fig. 9.13 Collection of data points showing torque/speed/force

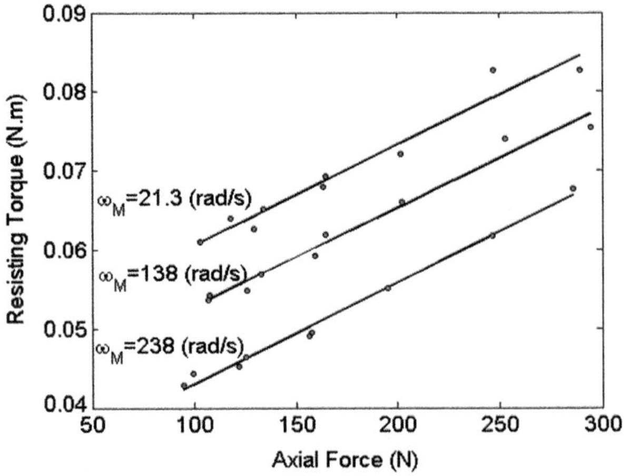

Fig 9.14 Sample measurement results. Variation of motor torque with applied axial load at constant speeds. *Dots*: measurements, *solid line*: fitted line to the data points

Based on (9.17) and using values of T_{f0} from Table 9.3 and T_f from Table 9.4, the residual friction of the lead screw supports is found to be $T_0 \approx 0.01$ Nm. Based on the parameter values in Table 9.4 and (9.12), $\xi_{W0} \approx 0.27$ was used in the calculations. Also using (9.16) and C_0 from Table 9.3 and C from Table 9.4, the damping coefficient of the end support is found to be $c \approx 4 \times 10^{-4}$ Nm s/rad. This value is adjusted in Sect. 9.6, since the system's stability (in simulations) depends heavily on the damping of the lead screw, and in the experimental results, there was quite a bit of variability.

Using (9.26), (9.27), and (9.28) and their identified values in Table 9.4, the three friction parameters defined by (9.24) can be calculated. These values are listed in

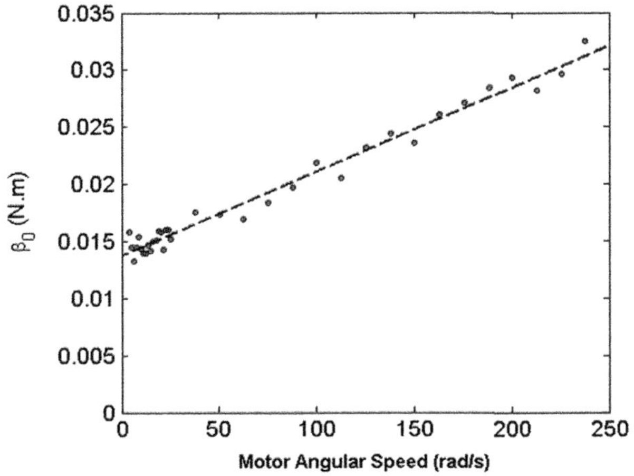

Fig. 9.15 Variation of β_0 with motor angular velocity

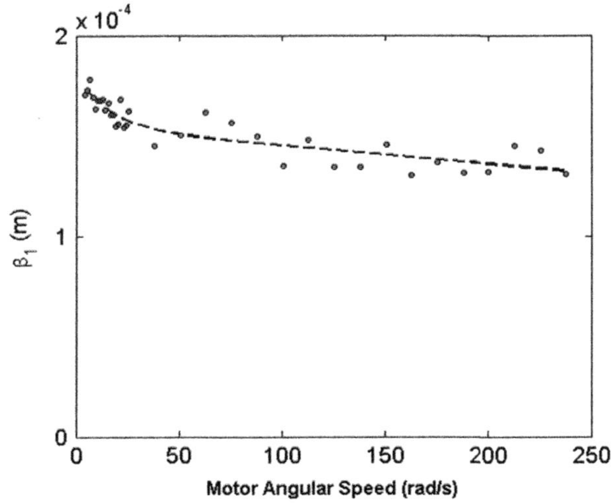

Fig. 9.16 Variation of β_1 with motor angular velocity

Table 9.5 and the resulting velocity-dependent coefficient of friction is plotted in Fig. 9.17.

9.6 Parameter Identification Step 2: Stiffness and Fine-Tuning

The identification formulation in the Sect. 9.3 depends on the knowledge of the sliding coefficient of friction of the gearbox (μ_{WG}) through the appearance of ξ_{W0} in (9.16), (9.17), (9.26), (9.27), and (9.28). Uncertainty in the value of this parameter,

Table 9.4 Identified parameters

Parameter	Value	Unit
T_f	0.0146	N m
C	7.34e−005	N m s/rad
r_0	0.38	s/rad
γ_0	1.54e−003	m
γ_1	2.59e−005	m
γ_2	−8.99e−008	m

Table 9.5 Numerical values of the identified parameters

Parameter	Value	Unit
μ_1	2.18e−1	–
μ_2	2.03e−2	–
μ_3	−4.47e−4	s/rad

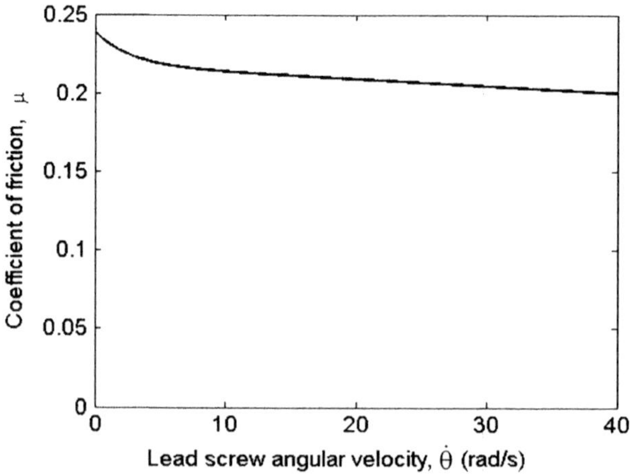

Fig. 9.17 Identified velocity-dependent coefficient of friction

together with the unknown nonlinearity of the coupling stiffness, necessitates a further step of parameter identification and fine-tuning. The approach in Sect. 9.3 was based on steady-state (no vibration) conditions. The idea of this step, however, is to use the open-loop lead screw response measurements where periodic motion (limit cycle) is observed. A set of carefully selected parameters which includes the overall torsional stiffness of the lead screw drive are adjusted to match the response of the mathematical model with the measurements. Only those input conditions (motor speed, ω_M and axial force, R) are considered in this step of identification where the system exhibits near steady-state periodic vibrations.

Figure 9.18a shows the results of a sample open-loop test where lead screw exhibited friction induced vibration. It is interesting to notice that, unlike the lead screw angular velocity, motor angular velocity does not exhibit considerable fluctuations. This is due to the high gear ratio between the DC motor and the lead screw. In the close-up view, Fig. 9.18b, the stick-slip rotational vibration of the lead screw can be clearly seen. During the same time, DC motor angular velocity remained almost constant.

Benefiting from the high gear ratio between the DC motor and the lead screw and the negligible moment of inertia of worm, worm gear, and coupling elements, the 3-DOF model in Fig. 9.8 is approximated by the 1-DOF model shown in Fig. 9.19.

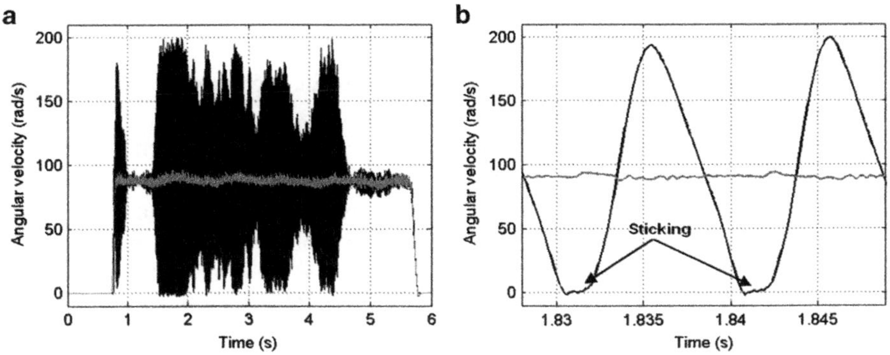

Fig. 9.18 (**a**) A sample of test results showing stick-slip in open-loop tests, (**b**) zoomed view. *Black*: lead screw angular velocity; *gray*: DC motor angular velocity

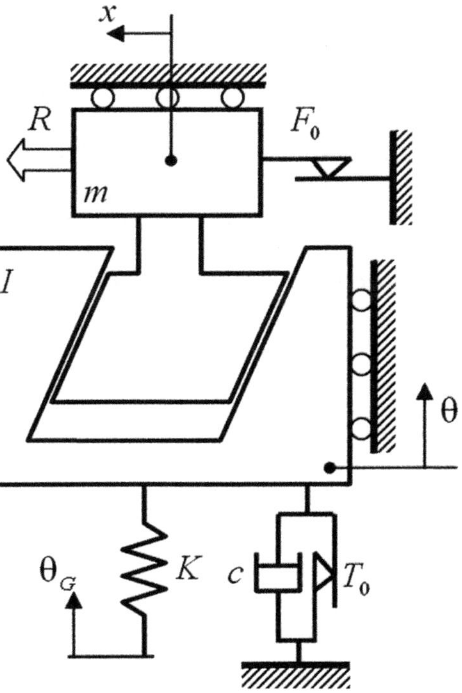

Fig. 9.19 Simplified 1-DOF system model

In this new and simplified model, the input is the gear angular displacement, θ_{G}, which is equal to $\alpha_W \theta_M$. Furthermore, from this point on, in the numerical simulations, it is assumed that the angular velocity output of gearbox, $\dot{\theta}_{\mathrm{G}} = \omega_{\mathrm{G}}$, is constant. The equation of motion of this system is given by

$$\left(I - \tan\lambda\xi m r_{\mathrm{m}}^2\right)\ddot{\theta} + K\theta + c\dot{\theta} = K\theta_{\mathrm{G}} - r_{\mathrm{m}}\xi\left(R - F_0 \mathrm{sgn}\left(\dot{\theta}\right)\right) - T_0 \mathrm{sgn}\left(\dot{\theta}\right), \tag{9.31}$$

where compared to (9.6) the only difference is in replacing k with K which represents the overall torsional stiffness of the drive.

As mentioned above, in the second step of the parameter identification, the response from the mathematical model, (9.31), is matched to the measurements by adjusting a set of parameters. The matching is only performed in the instances where a (possibly short lived) limit cycle behavior is observed in the measurements. The use of (almost) periodic data enables us to reduce the effects of transients thus improving the accuracy of the matching process. Consider the following cost function

$$\Psi(K, c, s_\mu, r_1) = \frac{\sum_{i=1}^{n}\left(\dot{\theta}^{\mathrm{M}}(i) - \dot{\theta}^{\mathrm{S}}(i)\right)^2}{\left(\max_i \dot{\theta}^{\mathrm{M}}(i)\right)^2} + \frac{\sum_{i=1}^{n}\left(\tilde{\theta}^{\mathrm{M}}(i) - \tilde{\theta}^{\mathrm{S}}(i)\right)^2}{\left(\max_i \tilde{\theta}^{\mathrm{M}}(i)\right)^2}, \tag{9.32}$$

which calculates the weighted sum of the squared differences between measured (superscript M) and simulated (superscript S) angular displacement ($\tilde{\theta}^{\mathrm{M}}$,$\tilde{\theta}^{\mathrm{S}}$) and angular velocity ($\dot{\theta}^{\mathrm{M}}$,$\dot{\theta}^{\mathrm{S}}$). The "$\sim$" signifies that the mean value is removed. Also, n is the number of data points included in the time window during which the measured system trajectory follows a limit-cycle.

The new parameters in the cost function Ψ in (9.32) are defined by the following modified velocity-dependent coefficient of friction

$$\mu = s_\mu\left(\mu_1 + \mu_2 \mathrm{e}^{-r_0|\dot{\theta}|} + \mu_3|\dot{\theta}|\right)\left(1 - \mathrm{e}^{-r_1|\dot{\theta}|}\right), \tag{9.33}$$

where s_μ is a scaling added to the identified friction to account for any variations in μ_{WG} from one experiment to the next. In addition, the friction coefficient is smoothed over near zero relative velocities to facilitate numerical integration and improve conformity of the simulation results to the test data[9] as the trajectories

[9]See results in Sect. 9.7.

approach the zero sliding velocity boundary. The parameters minimizing (9.33) are found numerically for every portion of the measured data where a limit cycle is found. These results are then used to obtain variations of each of the *fine-tuning* parameters in (9.33) (i.e., k, c, s_{μ}, and r_1) with respect to motor angular velocity, ω_M, and applied axial force, R.

9.7 Experimental Results: Step 2

In the second part of the experiments with the test setup shown in Fig. 9.9, the same configuration was used but without the motor speed controller. The motor servo amplifier was switched to "voltage mode" and the dSpace system was only used to collect data. A sample of the open-loop experimental results was shown in Fig. 9.18. Figure 9.20 shows the changes in the vibration amplitude as the applied axial force and the input angular velocity of lead screw is changed.

Each point in Fig. 9.20 represents the averaged experimental values of the amplitude of vibrations over a 2-ms interval, where a limit cycle was detected. These results show that the amplitude of vibration increases with gearbox output angular velocity. This finding also agrees with the perceived intensity of the audible noise generated by the system.

In Figs. 9.21–9.26, sample results are shown that compare measurements with the simulation results obtained from the 1-DOF model of Sect. 6.6 using the fine-tuned parameters. These results show the effectiveness of the fine-tuning step in matching the dynamical behavior of the model with that of the real system.

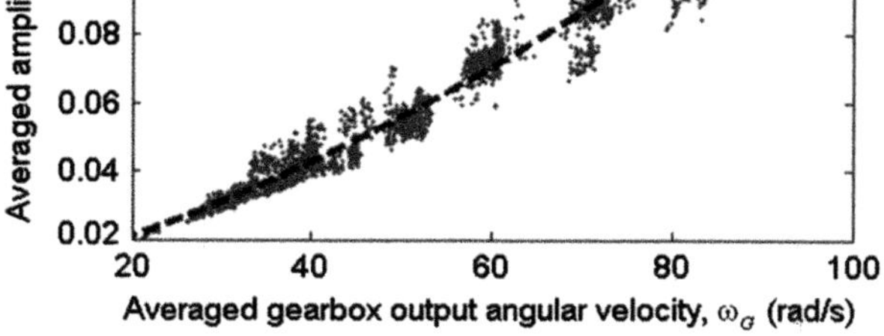

Fig. 9.20 Experimentally obtained variation of limit cycle vibration amplitude with input angular velocity (gearbox output) and axial force

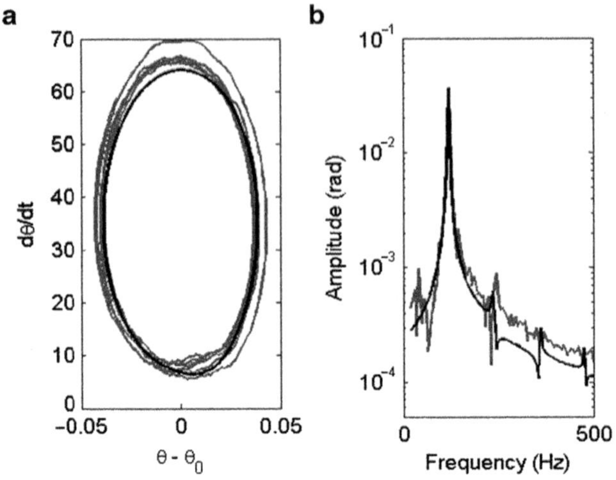

Fig. 9.21 Measurement vs. simulation example. Inputs: $R = 152.9$ (N), $\omega_G = 35.6$ (rad/s) – Parameters: $k = 1.18$ (Nm/rad), $c = 1.86 \times 10^{-4}$(Nm s/rad). (**a**) Phase plot and (**b**) frequency response. *Gray*: measurements, *black*: simulation

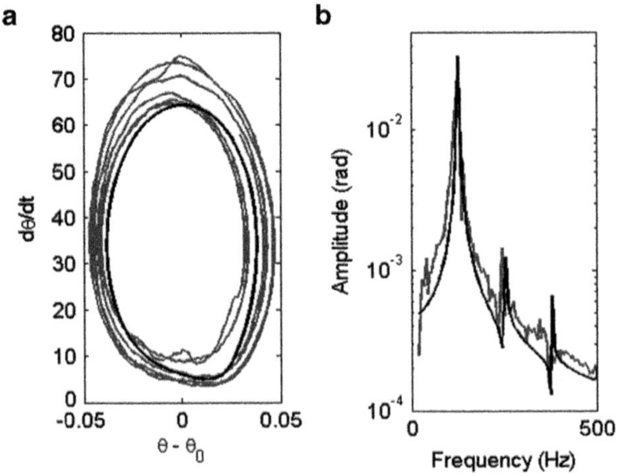

Fig. 9.22 Measurement vs. simulation example. Inputs: $R = 273.9$ (N), $\omega_G = 34.3$ (rad/s) – Parameters: $k = 1.31$ (Nm/rad), $c = 3.37 \times 10^{-4}$(N m s/rad). (**a**) Phase plot and (**b**) frequency response. *Gray*: measurements, *black*: simulation

In order to perform parameter studies through simulation, two-variable bilinear fitting is performed for each of the four fine-tuning parameters in (9.32) with respect to the gearbox output angular velocity, ω_G, and the applied axial force, R. Contour plots in Figs. 9.27–9.30 show the results of these data fittings.

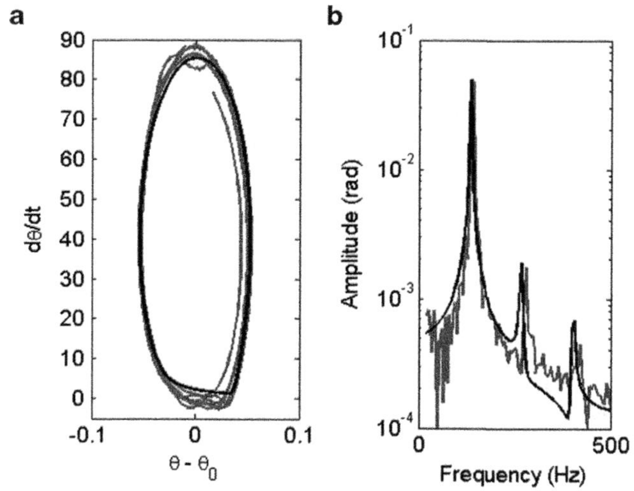

Fig. 9.23 Measurement vs. simulation example. Inputs: $R = 314.3$ (N), $\omega_G = 40.3$ (rad/s) – Parameters: $k = 1.65$ (Nm/rad), $c = 3.04 \times 10^{-4}$(Nms/rad). (**a**) Phase plot and (**b**) frequency response. *Gray*: measurements, *black*: simulation

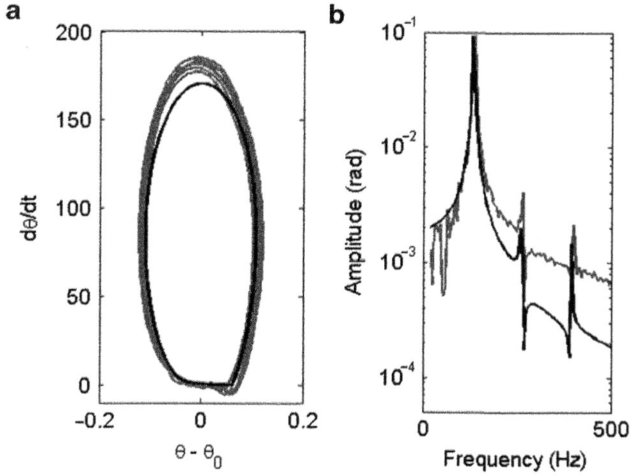

Fig. 9.24 Measurement vs. simulation example. Inputs: $R = 327.5$ (N), $\omega_G = 83.7$ (rad/s) – Parameters: $k = 1.75$ (Nm/rad), $c = 4.07 \times 10^{-4}$(Nms/rad). (**a**) Phase plot and (**b**) frequency response. *Gray*: measurements, *black*: simulation

It is interesting to note that, as expected, coupling stiffness (Fig. 9.27) varies mainly with the axial force and exhibits a "work-hardening" behavior. Also, the friction scaling (Fig. 9.30) was found to be very close to one which shows that the initial estimate of the value of μ_{WG} was quite accurate.

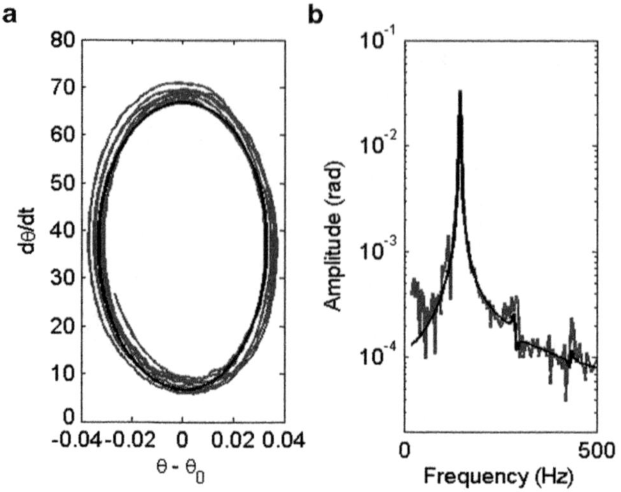

Fig. 9.25 Measurement vs. simulation example. Inputs: $R = 330.8$ (N), $\omega_G = 37.3$ (rad/s) – Parameters: $k = 1.88$ (Nm/rad), $c = 5.79 \times 10^{-4}$(Nms/rad). (**a**) Phase plot and (**b**) frequency response. *Gray*: measurements, *black*: simulation

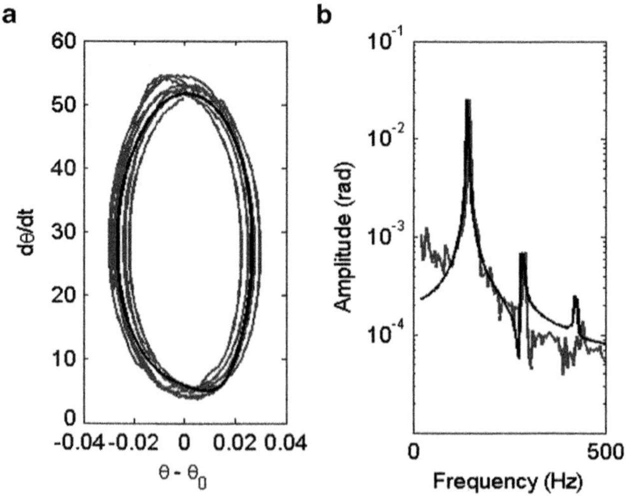

Fig. 9.26 Measurement vs. simulation example. Inputs: $R = 336.7$ (N), $\omega_G = 27.9$ (rad/s) – Parameters: $k = 1.67$ (Nm/rad), $c = 4.00 \times 10^{-4}$(Nms/rad). (**a**) Phase plot and (**b**) frequency response. *Gray*: measurements, *black*: simulation

9.8 Parameter Studies

Using the identified and fine-tuned model parameters, various performance maps can be obtained to study the effects of the variation of system parameters on the initiation of the limit cycles and the amplitude of the steady-state vibrations. The

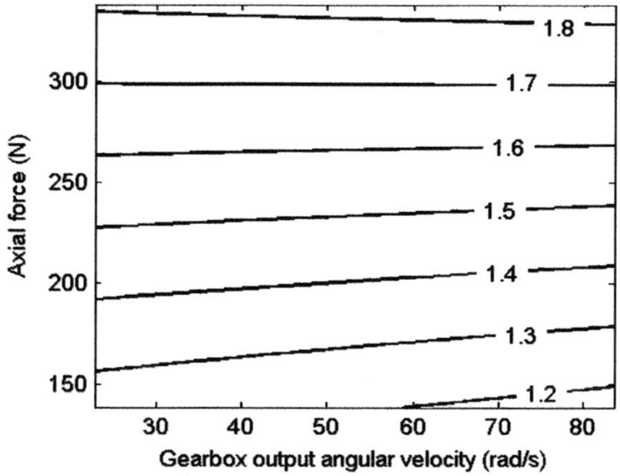

Fig. 9.27 Variation of coupling stiffness, $\boldsymbol{K}$, with gearbox output velocity and axial force

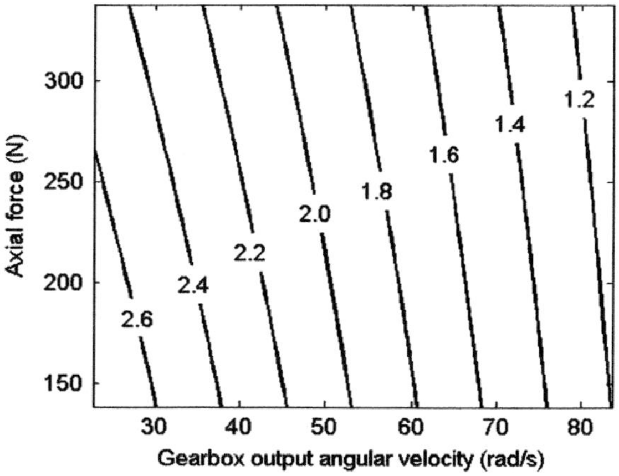

Fig. 9.28 Variation of friction boundary effect, $1/r_1$, with gearbox output velocity and axial force

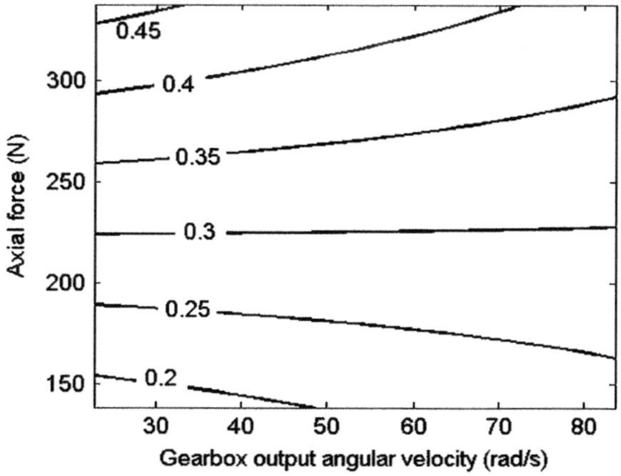

Fig. 9.29 Variation of lead screw support damping, $c\ (\times 10^{-3})$, with gearbox output velocity and axial force

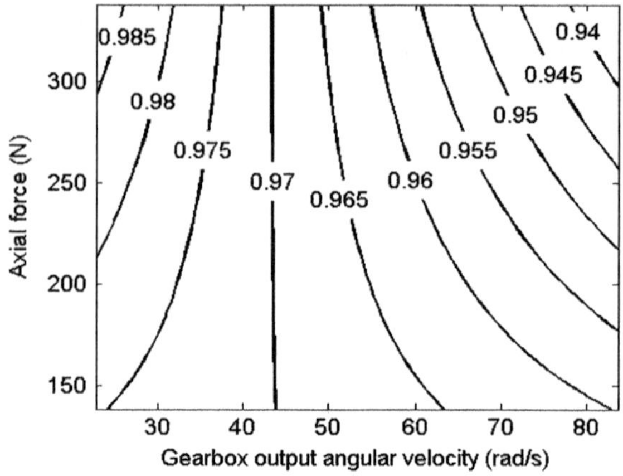

Fig. 9.30 Variation of friction scaling, s_μ, with gearbox output velocity and axial force

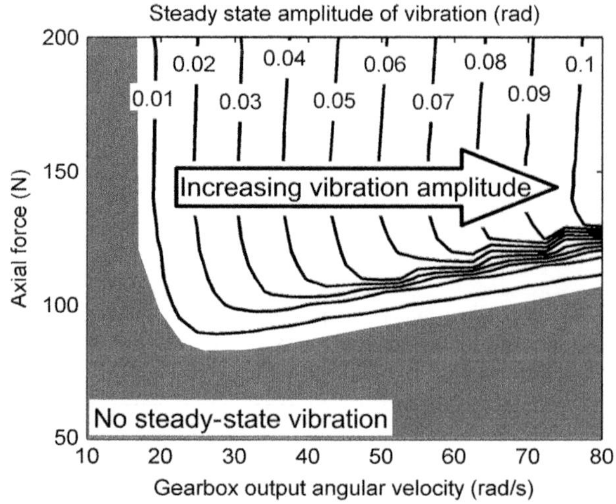

Fig. 9.31 Contour plots of the steady-state vibration amplitude vs. applied axial force and gearbox output speed

amplitude of vibrations is directly related to the generated audible noise from the system. In the following, the effects of the input angular velocity, the damping of the lead screw supports, and the coupling stiffness are investigated numerically.

9.8.1 *Effect of Input Angular Velocity*

Figure 9.31 shows the contours of the steady-state vibration amplitudes as a function of the gearbox output angular velocity and the applied axial force. It can be seen from this figure that, beyond a certain value of the applied axial force,

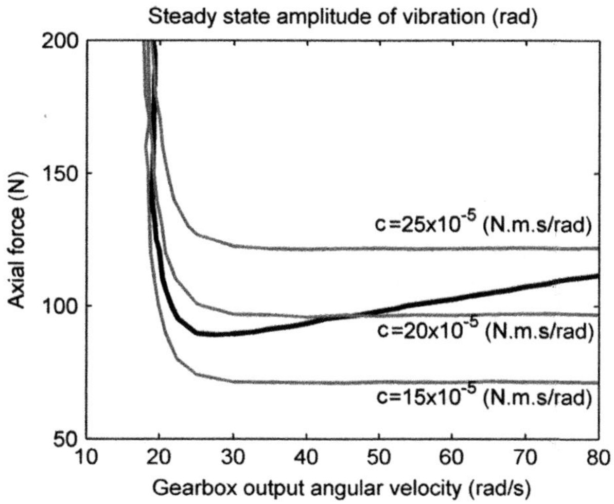

Fig. 9.32 Effect of lead screw rotational damping on the threshold of instabilities. The *thick black line* corresponds to the instability threshold in Fig. 9.31

increasing the angular velocity increases the amplitude of vibration. This finding is well correlated with the subjective tests on the audible noise intensity levels from the lead screw system and the experimental results in Fig. 9.20.

9.8.2 Effect of Damping

To investigate the effect of damping on the threshold of instabilities, the above simulations were repeated for three values of the constant damping coefficient. Figure 9.32 shows the result of these simulations where a vibration amplitude of 0.01 (rad) has been taken as the approximate threshold of stable/unstable regions. It can be seen that, by increasing the damping, instabilities occur at higher levels of axial force.[10]

9.8.3 Effect of Stiffness

The effect of the torsional stiffness of the coupling was also considered. Figure 9.33 shows the variation of the vibration amplitude as a function of the applied axial force and the torsional stiffness of the drive. This map was obtained using

[10]Refer to the results and discussions of Sect. 6.2.

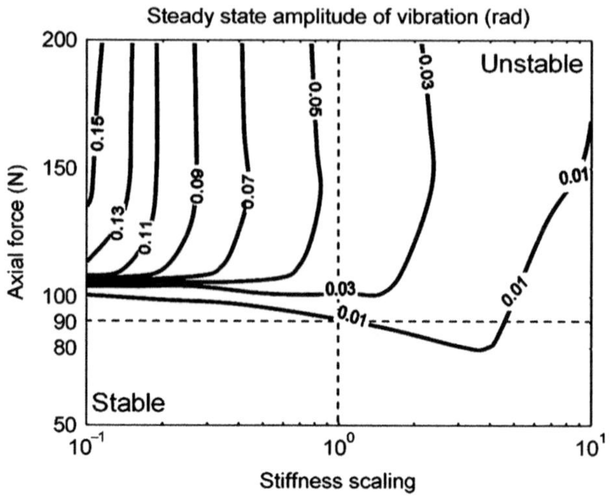

Fig. 9.33 The effects of coupling stiffness and axial loading on the dynamic behavior of the lead screw – Gearbox output angular velocity 40 (rad/s)

numerical simulations assuming $\omega_G = 40\ (\mathrm{rad/s})$. Similar plots can be obtained for other preset angular velocities.

Figure 9.33 shows that if the applied force is below 100 N, by either increasing *or* decreasing the stiffness of the drive from its current design value, rotational vibration leading to excessive noise may be eliminated. Note that the horizontal axis is a scaling parameter applied to the torsional stiffness of the drive. At higher axial loads, by using stiffer couplings – beyond ten times the current value – much lower vibration amplitudes are obtained. However, other design requirements may prevent the use of a high stiffness coupling in the system.

9.9 Conclusions

In this chapter, a two-step identification approach is developed to estimate various parameters of a lead screw drive system. In the first step, using the steady-sliding test results, the velocity effects (i.e., damping and velocity-dependent parts of friction) were separated from the force effects (i.e., coulomb coefficient of friction) and appropriate parameters were estimated using the least squares technique.

The identified velocity-dependent coefficient of friction was found to be a decreasing function of the sliding velocity in the lead screw drive mechanism of the powered seat adjuster. According to the results of Chap. 6, if damping is not sufficient, the negative damping effect caused by the negative gradient of coefficient of friction destabilizes the steady-sliding state and stick-slip type vibrations occur. This phenomenon was clearly observed in the open-loop tests performed on the lead screw mechanism (see Fig. 9.18).

The vibratory behavior of the system was utilized in the second step of the parameter identification. At instances where a limit cycles was observed, effective load-dependent torsional stiffness of the lead screw system was estimated by minimizing a cost function that quantified the difference between measured and numerically simulated displacements and velocities. Moreover, friction and damping parameters were adjusted so that maximum conformity between measurements and simulation results was achieved. The presented numerical simulations showed the accuracy of the identified mathematical model of the lead screw system under a wide range axial loading and input speed settings.

Parameter studies were performed to assess the effects of the lead screw rotational damping and the torsional stiffness of the drive on the onset of instabilities. Simulation results showed that by increasing the damping, instabilities occur at higher levels of applied force. In addition, it was shown that the torsional stiffness of the drive could change the axial loading range (dependent on the input speed) where the system generates significant levels of audible noise.

Appendices

Appendix A: First-Order Averaging Theorem

Consider the following system in standard form

$$\dot{\mathbf{x}} = \varepsilon \mathbf{f}(t, \mathrm{x}, \varepsilon), \quad \mathbf{x}(0) = \mathbf{x}_0, \tag{A.1}$$

Suppose

- *The function* $\mathbf{f} : R_+ \times D \times [0, \varepsilon_0] \to R^n$ *is T-periodic with respect to t for* $x \in D$. $D \subset R^n$ *is an open bounded set and* $\varepsilon_0 > 0$ *is some number.*
- *There exists a constant* $M > 0$ *such that* $\|\mathbf{f}(t, \mathbf{x}, \varepsilon)\| \leq M$.
- $\mathbf{f}(t, \mathbf{x}, \varepsilon)$ *is Lipschitz continuous with respect to* $\mathbf{x}$ *and* ε *with Lipschitz constants* λ_x *and* λ_ε, *respectively.*
- *The average* $\bar{\mathbf{f}}(\mathbf{x}) = 1/T \int_0^T \mathbf{f}(\tau, \mathbf{x}, 0) \mathrm{d}\tau$ *exists uniformly with respect to* $\mathbf{x}$.
- *Consider the averaged system;*

$$\dot{\mathbf{z}} = \varepsilon \bar{\mathbf{f}}(\mathbf{z}), \quad \mathbf{z}(0) = \mathbf{x}_0. \tag{A.2}$$

- *The solution of (A.2),* $\mathbf{z}(t; 0, \mathbf{x}_0)$, *belongs to the interior subset of D on time scale* $1/\varepsilon$.

Then, there exists $c > 0, \varepsilon_0 > 0$, *and* $L > 0$, *such that the following holds for the solutions of (A.1) and (A.2)*

$$\|\mathbf{x}(t, \varepsilon) - \mathbf{z}(t, \varepsilon)\| \leq c\varepsilon. \tag{A.3}$$

For $0 \leq \varepsilon \leq \varepsilon_0$ *and* $0 \leq t \leq L/\varepsilon$. *Also, c is independent of* ε.

Proof. *We follow the steps given in [56, 57]. Let*

$$\mathbf{E}(t, \varepsilon) = \mathbf{x}(t, \varepsilon) - \mathbf{z}(t, \varepsilon), \tag{A.4}$$

denote the error. From the two differential equations, (A.4) is found as

$$\mathbf{E}(t,\varepsilon) = \varepsilon \int_0^t \left[\mathbf{f}(\tau, \mathbf{x}(\tau,\varepsilon), \varepsilon) - \bar{\mathbf{f}}(\mathbf{z}(\tau,\varepsilon))\right] \mathrm{d}\tau. \tag{A.5}$$

The integrant in (A.5) can be written as

$$[\mathbf{f}(\tau, \mathbf{x}, \varepsilon) - \mathbf{f}(\tau, \mathbf{z}, \varepsilon)] + [\mathbf{f}(\tau, \mathbf{z}, \varepsilon) - \mathbf{f}(\tau, \mathbf{z}, 0)] + \left[\mathbf{f}(\tau, \mathbf{z}, 0) - \bar{\mathbf{f}}(\mathbf{z})\right],$$

where the arguments of **E**, **x**, and **z** are omitted for brevity. As a result, from (A.5) we have

$$\begin{aligned}\|\mathbf{E}\| \leq \varepsilon \int_0^t \|\mathbf{f}(\tau, \mathbf{x}, \varepsilon) - \mathbf{f}(\tau, \mathbf{z}, \varepsilon)\| \mathrm{d}\tau + \varepsilon \int_0^t \|\mathbf{f}(\tau, \mathbf{z}, \varepsilon) - \mathbf{f}(\tau, \mathbf{z}, 0)\| \mathrm{d}\tau \\ + \varepsilon \left\| \int_0^t \left[\mathbf{f}(\tau, \mathbf{z}, 0) - \bar{\mathbf{f}}(\mathbf{z})\right] \mathrm{d}\tau \right\|.\end{aligned} \tag{A.6}$$

The first and second terms on the right-hand side of (A.6) can be estimated using the Lipschitz constants λ_x and λ_ε

$$\|\mathbf{E}\| \leq \varepsilon \lambda_x \int_0^t \|\mathbf{E}\| \mathrm{d}\tau + \varepsilon^2 \lambda_\varepsilon t + \varepsilon \left\| \int_0^t \left[\mathbf{f}(\tau, \mathbf{z}, 0) - \bar{\mathbf{f}}(\mathbf{z})\right] \mathrm{d}\tau \right\|. \tag{A.7}$$

The third term in (A.6) or (A.7) is estimated as follows:

$$\begin{aligned}\left\| \int_0^t \left[\mathbf{f}(\tau, \mathbf{z}, 0) - \bar{\mathbf{f}}(\mathbf{z})\right] \mathrm{d}\tau \right\| \leq \sum_{i=1}^{N} \left\| \int_{(i-1)T}^{iT} \left[\mathbf{f}(\tau, \mathbf{z}, 0) - \bar{\mathbf{f}}(\mathbf{z})\right] \mathrm{d}\tau \right\| \\ + \left\| \int_{NT}^{t} \left[\mathbf{f}(\tau, \mathbf{z}, 0) - \bar{\mathbf{f}}(\mathbf{z})\right] \mathrm{d}\tau \right\|,\end{aligned} \tag{A.8}$$

where N is chosen such that, $NT \leq t \leq (N+1)T$. We have

$$\begin{aligned}&\int_{(i-1)T}^{iT} \left[\mathbf{f}(\tau, \mathbf{z}(\tau,\varepsilon), 0) - \bar{\mathbf{f}}(\mathbf{z}(\tau,\varepsilon))\right] \mathrm{d}\tau \\ &= \int_{(i-1)T}^{iT} [\mathbf{f}(\tau, \mathbf{z}(\tau,\varepsilon), 0) - \bar{\mathbf{f}}(\mathbf{z}(\tau,\varepsilon)) - \mathbf{f}((i-1)T, \\ &\quad \times \mathbf{z}((i-1)T,\varepsilon), 0) + \bar{\mathbf{f}}(\mathbf{z}((i-1)T,\varepsilon))] \mathrm{d}\tau\end{aligned}$$

which holds since $\int_{(i-1)T}^{iT} \left[\mathbf{f}(\tau, \mathbf{z}((i-1)T,\varepsilon), 0) - \bar{\mathbf{f}}(\mathbf{z}((i-1)T,\varepsilon))\right] \mathrm{d}\tau = 0$. As a result

$$
\begin{aligned}
&\sum_{i=1}^{N}\left\|\int_{(i-1)T}^{iT}\left[\mathbf{f}(\tau,\mathbf{z},0)-\bar{\mathbf{f}}(\mathbf{z})\right]\mathrm{d}\tau\right\| \\
&=\sum_{i=1}^{N}\left\|\int_{(i-1)T}^{iT}\left[\mathbf{f}(\tau,\mathbf{z}(\tau,\varepsilon),0)-\bar{\mathbf{f}}(\mathbf{z}(\tau,\varepsilon))-\mathbf{f}(\tau,\mathbf{z}((i-1)T,\varepsilon),0)+\bar{\mathbf{f}}(\mathbf{z}((i-1)T,\varepsilon))\right]\mathrm{d}\tau\right\|, \\
&\leq\sum_{i=1}^{N}\left\|\int_{(i-1)T}^{iT}[\mathbf{f}(\tau,\mathbf{z}(\tau,\varepsilon),0)-\mathbf{f}(\tau,\mathbf{z}((i-1)T,\varepsilon),0)]\mathrm{d}\tau\right\| \\
&+\sum_{i=1}^{N}\left\|\int_{(i-1)T}^{iT}\left[\bar{\mathbf{f}}(\mathbf{z}((i-1)T,\varepsilon))-\bar{\mathbf{f}}(\mathbf{z}(\tau,\varepsilon))\right]\mathrm{d}\tau\right\|, \\
&\leq 2\lambda_x\sum_{i=1}^{N}\left\|\int_{(i-1)T}^{iT}[\mathbf{z}(\tau,\varepsilon)-\mathbf{z}((i-1)T,\varepsilon)]\mathrm{d}\tau\right\|, \\
&\leq 2\varepsilon\lambda_x T^2 MN. && \text{(A.9)}
\end{aligned}
$$

The first inequality in (A.9) holds since $\bar{\mathbf{f}}(\mathbf{x})$ has the same Lipschitz constant as $\mathbf{f}(t,\mathbf{x},\varepsilon)$ with respect to $\mathbf{x}$:

$$
\begin{aligned}
\left\|\bar{\mathbf{f}}(\mathbf{x}_1)-\bar{\mathbf{f}}(\mathbf{x}_2)\right\| &= \frac{1}{T}\left\|\int_0^T[\mathbf{f}(\tau,\mathbf{x}_1,0)-\mathbf{f}(\tau,\mathbf{x}_2,0)]\mathrm{d}\tau\right\| \\
&\leq\frac{1}{T}\int_0^T\|\mathbf{f}(\tau,\mathbf{x}_1,0)-\mathbf{f}(\tau,\mathbf{x}_2,0)\|\mathrm{d}\tau, \\
&\leq\frac{\lambda_x}{T}\int_0^T\|\mathbf{x}_1-\mathbf{x}_2\|\mathrm{d}\tau=\lambda_x\|\mathbf{x}_1-\mathbf{x}_2\|.
\end{aligned}
$$

The second inequality in (A.9) holds since $\mathbf{z}(\tau,\varepsilon)$ is the solution of (A.2) and as such it is slowly varying:

$$\|\mathbf{z}(\tau,\varepsilon)-\mathbf{z}((i-1)T,\varepsilon)\|\leq\varepsilon TM.$$

Also note that

$$\left\|\int_{NT}^{t}\left[\mathbf{f}(\tau,\mathbf{z},0)-\bar{\mathbf{f}}(\mathbf{z})\right]\mathrm{d}\tau\right\|\leq 2TM. \tag{A.10}$$

Using (A.9) and (A.10), (A.8) becomes

$$\left\|\int_0^t\left[\mathbf{f}(\tau,\mathbf{z},0)-\bar{\mathbf{f}}(\mathbf{z})\right]\mathrm{d}\tau\right\|\leq 2\varepsilon T^2M\lambda_xN+2TM\leq 2TM(\lambda_xL+1). \tag{A.11}$$

Note that in (A.11), the inequality $NT\leq t\leq L/\varepsilon$ was used. Finally, using (A.11) the error estimate (A.7) becomes

$$\|\mathbf{E}\| \le \varepsilon\lambda_x \int_0^t \|\mathbf{E}\| \mathrm{d}\tau + \varepsilon\lambda_\varepsilon L + 2\varepsilon TM(\lambda_x L + 1). \tag{A.12}$$

Applying the Gronwall's lemma [56, 57] to (A.12) yields

$$\begin{aligned} \|\mathbf{x}(t,\varepsilon) - \mathbf{z}(t,\varepsilon)\| &\le \varepsilon[\lambda_\varepsilon L + 2TM(\lambda_x L + 1)]\mathrm{e}^{\varepsilon\lambda_x t} \\ &\le \varepsilon[\lambda_\varepsilon L + 2TM(\lambda_x L + 1)]\mathrm{e}^{\lambda_x L}. \end{aligned} \tag{A.13}$$

Taking $c = [\lambda_\varepsilon L + 2TM(\lambda_x L + 1)]\mathrm{e}^{\lambda_x L}$ completes the proof. ■

Appendix B: Application of Higher-Order Averaging

The accuracy of the results obtained in Sect. 6.3 depends heavily on the "size" of the system parameters. In practical applications, such as the experimental example of Chap. 9, the $O(\varepsilon)$ error of the first-order averaging may not be sufficiently accurate for the entire range of parameters in the domain of interest. A possible way to improve the accuracy of the amplitude and frequency estimates is to extend the averaging to higher orders.

A slightly different approach to deriving the equation of motion (of the system treated in Chap. 6) in standard form is used here. In Sect. B.1, the approximate system's equation accurate to $O(\varepsilon^4)$ is derived. The steps needed to carry out the averaging process up to the third order are presented in Sect. B.2. In Sect. B.3, a numerical example is presented that compares averaging results with numerical simulation results and actual measurements.

B.1 Equation of Motion in Standard Form

Two simplifying assumptions are made in Chap. 6 to reach the averaged equation (6.41) – namely, the contact force does not change its sign and lead screw velocity does not change its sign. These assumptions are made here from the start to ensure that the equations have the required smoothness properties. For simplicity, the analysis is limited to the case of $R > 0$ (thus, $N > 0$) and $\omega > 0$(thus, $\dot{\theta} > 0$). We start from (6.21)

$$\begin{aligned} &v'' + \varepsilon\left[1 - \varepsilon\Xi_1(v, v', \varepsilon)\widehat{m}\right]^{-1}\widehat{c}v' + \left[1 - \varepsilon\Xi_1(v, v', \varepsilon)\widehat{m}\right]^{-1}v \\ &\quad = \varepsilon\widehat{R}\left[1 - \varepsilon\Xi_1(v, v', \varepsilon)\widehat{m}\right]^{-1}\left[\Xi_0(\varepsilon) - \Xi_1(v, v', \varepsilon)\right], \end{aligned} \tag{B.1}$$

where $\Xi_0(\varepsilon)$ and $\Xi_1(v, v', \varepsilon)$ are given by (6.22) and (6.23), respectively. The above assumptions simplify these functions to

$$\Xi_0(\varepsilon) = \frac{\widehat{\mu}_0 - 1}{1 + \varepsilon^2 \widehat{\mu}_0}, \tag{B.2}$$

$$\Xi_1(v', \varepsilon) = \frac{\widehat{\mu}(v') - 1}{1 + \varepsilon^2 \widehat{\mu}(v')}, \tag{B.3}$$

where $\widehat{\mu}(v')$ is obtained from (6.24)

$$\widehat{\mu}(v') = \left(\hat{\mu}_1 + \hat{\mu}_2 \mathrm{e}^{-r_0 \Omega(v'+1)}\right)\left(1 - \mathrm{e}^{-r_1 \Omega(v'+1)}\right) + \hat{\mu}_3 \Omega(v' + 1), \tag{B.4}$$

$\Xi_0(\varepsilon)$ and $\Xi_1(v', \varepsilon)$ can be expanded in powers of ε as

$$\Xi_0(\varepsilon) = \left(\widehat{\mu}_0 - 1\right)\left(1 - \varepsilon^2 \widehat{\mu}_0^2\right) + O(\varepsilon^4), \tag{B.5}$$

$$\Xi_1(v', \varepsilon) = \left(\widehat{\mu} - 1\right)\left(1 - \varepsilon^2 \widehat{\mu}^2\right) + O(\varepsilon^4). \tag{B.6}$$

Also

$$\begin{aligned}\left(1 - \varepsilon \frac{\widehat{\mu} - 1}{1 + \varepsilon^2 \widehat{\mu}} \widehat{m}\right)^{-1} &= 1 + \varepsilon\left(\widehat{\mu} - 1\right)\widehat{m} + \varepsilon^2\left(\widehat{\mu} - 1\right)^2 \widehat{m}^2 \\ &\quad + \varepsilon^3\left(\widehat{\mu} - 1\right)^3 \widehat{m}^3 - \varepsilon^3 \widehat{\mu}\left(\widehat{\mu} - 1\right)\widehat{m} + O(\varepsilon^4).\end{aligned} \tag{B.7}$$

Substituting (B.5), (B.6), and (B.7) into (B.1) and keeping terms up to $O(\varepsilon^3)$ gives

$$v'' + v = \varepsilon f_1(v, v') + \varepsilon^2 f_2(v, v') + \varepsilon^3 f_3(v, v'), \tag{B.8}$$

where

$$f_1(v, v') = -\widehat{c}v' - \left(\widehat{\mu} - 1\right)\widehat{m}v + \widehat{R}\left(\widehat{\mu}_0 - \widehat{\mu}\right), \tag{B.9}$$

$$f_2(v, v') = -\left(\widehat{\mu} - 1\right)\widehat{m}\widehat{c}v' - \left(\widehat{\mu} - 1\right)^2 \widehat{m}^2 v + \widehat{R}\left(\widehat{\mu} - 1\right)\left(\widehat{\mu}_0 - \widehat{\mu}\right)\widehat{m}, \tag{B.10}$$

$$\begin{aligned}f_3(v, v') = &-\left(\widehat{\mu} - 1\right)^2 \widehat{m}^2 \widehat{c}v' - \left(\widehat{\mu} - 1\right)^3 \widehat{m}^3 v + \widehat{R}\left(\widehat{\mu} - 1\right)^2\left(\widehat{\mu}_0 - \widehat{\mu}\right)\widehat{m}^2 \\ &+ \widehat{\mu}\left(\widehat{\mu} - 1\right)\widehat{m}v - \widehat{R}\left[\widehat{\mu}_0^2\left(\widehat{\mu}_0 - 1\right) - \widehat{\mu}^2\left(\widehat{\mu} - 1\right)\right],\end{aligned} \tag{B.11}$$

Unfortunately, the expressions involved are too cumbersome to be of any practical use in closed form. However, if approached numerically, these higher-order approximations can be used to estimate the amplitude of the steady-state vibrations, efficiently.

B.2 Higher-Order Averaging Formulation

In this appendix, following [55], the general formulations of first-, second-, and third-order averaging are derived for a system of differential equations in the standard form:

$$\dot{\mathbf{x}} = \varepsilon \mathbf{X}_1(t, \mathbf{x}) + \varepsilon^2 \mathbf{X}_2(t, \mathbf{x}) + \varepsilon^3 \mathbf{X}_3(t, \mathbf{x}), \tag{B.12}$$

where

$$\mathbf{x} = \begin{Bmatrix} a \\ \beta \end{Bmatrix}, \tag{B.13}$$

$$\mathbf{X}_i = \begin{Bmatrix} X_{i1} \\ X_{i2} \end{Bmatrix}, \tag{B.14}$$

To clarify the notations, the following expressions are given here

$$\nabla \mathbf{X}_i(t, \boldsymbol{\xi}) = \nabla_{\boldsymbol{\xi}} \mathbf{X}_i(t, \boldsymbol{\xi}) = \begin{bmatrix} \dfrac{\partial X_{i1}}{\partial a} & \dfrac{\partial X_{i1}}{\partial \beta} \\ \dfrac{\partial X_{i2}}{\partial a} & \dfrac{\partial X_{i2}}{\partial \beta} \end{bmatrix}, \tag{B.15}$$

and

$$\nabla^2 X_{ij}(t, \boldsymbol{\xi}) = \begin{bmatrix} \dfrac{\partial^2 X_{ij}}{\partial a^2} & \dfrac{\partial^2 X_{ij}}{\partial a \partial \beta} \\ \dfrac{\partial^2 X_{ij}}{\partial \beta \partial a} & \dfrac{\partial^2 X_{ij}}{\partial \beta^2} \end{bmatrix}. \tag{B.16}$$

B.2.1 First-Order Averaging

Introducing the following *near-identity transform*

$$\mathbf{x} = \boldsymbol{\xi} + \varepsilon \mathbf{F}_1(t, \boldsymbol{\xi}), \tag{B.17}$$

where $\boldsymbol{\xi}$ is the solution of

$$\dot{\boldsymbol{\xi}} = \varepsilon \mathbf{P}_1(\boldsymbol{\xi}), \tag{B.18}$$

where $\mathbf{F}_1$ and $\mathbf{P}_1$ are unknown functions to be determined. Substituting (B.17) into (B.12) (neglecting ε^2 and ε^3 terms)

$$\frac{\mathrm{d}}{\mathrm{d}t}(\boldsymbol{\xi} + \varepsilon \mathbf{F}_1(t, \boldsymbol{\xi})) = \varepsilon \mathbf{X}_1(t, \boldsymbol{\xi} + \varepsilon \mathbf{F}_1(t, \boldsymbol{\xi})),$$

expanding RHS using Taylor series expansion and substituting (B.18) gives

$$\varepsilon \mathbf{P}_1 + \varepsilon \frac{\partial \mathbf{F}_1}{\partial t} + \varepsilon^2 \nabla \mathbf{F}_1 \cdot \mathbf{P}_1 = \varepsilon \mathbf{X}_1 + \varepsilon^2 \nabla \mathbf{X}_1 \cdot \mathbf{F}_1.$$

Neglecting ε^2 terms

$$\frac{\partial \mathbf{F}_1(t, \boldsymbol{\xi})}{\partial t} = \mathbf{X}_1(t, \boldsymbol{\xi}) - \mathbf{P}_1(\boldsymbol{\xi}). \tag{B.19}$$

The solution to this equation can be written as

$$\mathbf{P}_1(\boldsymbol{\xi}) = \frac{1}{T} \int_0^T \mathbf{X}_1(t, \boldsymbol{\xi}) \mathrm{d}t. \tag{B.20}$$

Substituting (B.20) into (B.19) and integrating

$$\mathbf{F}_1(t, \boldsymbol{\xi}) = \int_0^t [\mathbf{X}_1(\tau, \boldsymbol{\xi}) - \mathbf{P}_1(\boldsymbol{\xi})] \mathrm{d}\tau + \mathbf{a}_1(\boldsymbol{\xi}), \tag{B.21}$$

where following [55], $\mathbf{a}_1(\boldsymbol{\xi})$ is chosen such that $\mathbf{F}_1(t, \boldsymbol{\xi})$ has a zero mean, i.e.,

$$\mathbf{a}_1(\boldsymbol{\xi}) = \frac{1}{T} \int_0^T \int_0^t [\mathbf{X}_1(\tau, \boldsymbol{\xi}) - \mathbf{P}_1(\boldsymbol{\xi})] \mathrm{d}\tau \mathrm{d}t. \tag{B.22}$$

It can be shown that the solution of (B.18) remains $O(\varepsilon)$ close to the solution of the original differential equation (B.12) on a time scale of $O(1/\varepsilon)$, i.e.,

$$\|\mathbf{x}(t) - \boldsymbol{\xi}(t)\| \leq k\varepsilon, \quad \forall 0 \leq t \leq \frac{L}{\varepsilon}, \tag{B.23}$$

for some $k > 0$ and $L > 0$.

B.2.2 Second-Order Averaging

The near-identity transform is modified to

$$\mathbf{x} = \boldsymbol{\xi} + \varepsilon \mathbf{F}_1(t, \boldsymbol{\xi}) + \varepsilon^2 \mathbf{F}_2(t, \boldsymbol{\xi}), \tag{B.24}$$

where $\boldsymbol{\xi}$ is the solution of

$$\dot{\boldsymbol{\xi}} = \varepsilon \mathbf{P}_1(\boldsymbol{\xi}) + \varepsilon^2 \mathbf{P}_2(\boldsymbol{\xi}), \tag{B.25}$$

where $\mathbf{F}_1$ and $\mathbf{P}_1$ are defined as before, and $\mathbf{F}_2$ and $\mathbf{P}_2$ are unknown functions to be determined. Substituting (B.24) into (B.12) (neglecting ε^3 terms)

$$\begin{aligned}\frac{\mathrm{d}}{\mathrm{d}t}\left(\boldsymbol{\xi} + \varepsilon \mathbf{F}_1(t,\boldsymbol{\xi}) + \varepsilon^2 \mathbf{F}_2(t,\boldsymbol{\xi})\right) &= \varepsilon \mathbf{X}_1\left(t, \boldsymbol{\xi} + \varepsilon \mathbf{F}_1(t,\boldsymbol{\xi}) + \varepsilon^2 \mathbf{F}_2(t,\boldsymbol{\xi})\right) \\ &\quad + \varepsilon^2 \mathbf{X}_2\left(t, \boldsymbol{\xi} + \varepsilon \mathbf{F}_1(t,\boldsymbol{\xi}) + \varepsilon^2 \mathbf{F}_2(t,\boldsymbol{\xi})\right).\end{aligned}$$

Expanding the RHS using Taylor series expansion

$$\begin{aligned}\dot{\boldsymbol{\xi}} + \varepsilon \frac{\partial \mathbf{F}_1}{\partial t} + \varepsilon \nabla \mathbf{F}_1 \cdot \dot{\boldsymbol{\xi}} + \varepsilon^2 \frac{\partial \mathbf{F}_2}{\partial t} + \varepsilon^2 \nabla \mathbf{F}_2 \cdot \dot{\boldsymbol{\xi}} &= \varepsilon \mathbf{X}_1 + \varepsilon^2 \nabla \mathbf{X}_1 \cdot \mathbf{F}_1 + \varepsilon^3 \nabla \mathbf{X}_1 \cdot \mathbf{F}_2 \\ &\quad + \varepsilon^3 \begin{bmatrix} [\mathbf{F}_1 + \varepsilon \mathbf{F}_2]^{\mathrm{T}} \nabla^2 X_{11} [\mathbf{F}_1 + \varepsilon \mathbf{F}_2] \\ [\mathbf{F}_1 + \varepsilon \mathbf{F}_2]^{\mathrm{T}} \nabla^2 X_{12} [\mathbf{F}_1 + \varepsilon \mathbf{F}_2] \end{bmatrix} \\ &\quad + \varepsilon^2 \mathbf{X}_2 + \varepsilon^3 \nabla \mathbf{X}_2 \cdot \mathbf{F}_1.\end{aligned}$$

Substituting (B.25) and neglecting ε^3 terms

$$\mathbf{P}_2 + \nabla \mathbf{F}_1 \cdot \mathbf{P}_1 + \frac{\partial \mathbf{F}_2}{\partial t} = \nabla \mathbf{X}_1 \cdot \mathbf{F}_1 + \mathbf{X}_2. \tag{B.26}$$

The solution to (B.26) can be written as

$$\mathbf{P}_2(\boldsymbol{\xi}) = \frac{1}{T} \int_0^T [\nabla \mathbf{X}_1(t,\boldsymbol{\xi}) \cdot \mathbf{F}_1(t,\boldsymbol{\xi}) - \nabla \mathbf{F}_1(t,\boldsymbol{\xi}) \cdot \mathbf{P}_1(\boldsymbol{\xi}) + \mathbf{X}_2(t,\boldsymbol{\xi})]\mathrm{d}t,$$

or since $\nabla \mathbf{F}_1(t,\boldsymbol{\xi}) \cdot \mathbf{P}_1(\boldsymbol{\xi})$ has a zero mean

$$\mathbf{P}_2 = \frac{1}{T} \int_0^T [\nabla \mathbf{X}_1(t,\boldsymbol{\xi}) \cdot \mathbf{F}_1(t,\boldsymbol{\xi}) + \mathbf{X}_2(t,\boldsymbol{\xi})]\mathrm{d}t, \tag{B.27}$$

where $\mathbf{F}_1(t,\boldsymbol{\xi})$ is given by (B.21). Subsequently, integrating (B.26) gives

$$\mathbf{F}_2(t,\boldsymbol{\xi}) = \int_0^t [\nabla \mathbf{X}_1(\tau,\boldsymbol{\xi}) \cdot \mathbf{F}_1(\tau,\boldsymbol{\xi}) - \nabla \mathbf{F}_1(\tau,\boldsymbol{\xi}) \cdot \mathbf{P}_1(\boldsymbol{\xi}) + \mathbf{X}_2(\tau,\boldsymbol{\xi}) - \mathbf{P}_2(\boldsymbol{\xi})]\mathrm{d}\tau + \mathbf{a}_2(\boldsymbol{\xi}), \tag{B.28}$$

where $\mathbf{a}_2(\boldsymbol{\xi})$ is chosen such that $\mathbf{F}_2(t,\boldsymbol{\xi})$ has zero mean, i.e.,

$$\mathbf{a}_2(\boldsymbol{\xi}) = \frac{1}{T} \int_0^T \int_0^t [\nabla \mathbf{X}_1(\tau,\boldsymbol{\xi}) \cdot \mathbf{F}_1(\tau,\boldsymbol{\xi}) - \nabla \mathbf{F}_1(\tau,\boldsymbol{\xi}) \cdot \mathbf{P}_1(\boldsymbol{\xi}) + \mathbf{X}_2(\tau,\boldsymbol{\xi}) - \mathbf{P}_2(\boldsymbol{\xi})]\mathrm{d}\tau \mathrm{d}t. \tag{B.29}$$

It can be shown that given $\boldsymbol{\xi}(t)$ to be the solution of (B.25),

$$\|\mathbf{x}(t) - \boldsymbol{\xi}(t) - \varepsilon \mathbf{F}_1(t, \boldsymbol{\xi}(t))\| = O(\varepsilon^2), \tag{B.30}$$

For the time scale of $O(1/\varepsilon)$.

B.2.3 Third-Order Averaging

Similar to the previous section, the near-identity transform is now defined as

$$\mathbf{x} = \boldsymbol{\xi} + \varepsilon \mathbf{F}_1(t, \boldsymbol{\xi}) + \varepsilon^2 \mathbf{F}_2(t, \boldsymbol{\xi}) + \varepsilon^3 \mathbf{F}_3(t, \boldsymbol{\xi}), \tag{B.31}$$

where $\boldsymbol{\xi}$ is the solution of

$$\dot{\boldsymbol{\xi}} = \varepsilon \mathbf{P}_1(\boldsymbol{\xi}) + \varepsilon^2 \mathbf{P}_2(\boldsymbol{\xi}) + \varepsilon^3 \mathbf{P}_3(\boldsymbol{\xi}), \tag{B.32}$$

where $\mathbf{F}_1$, $\mathbf{F}_2$ and $\mathbf{P}_1$, $\mathbf{P}_2$ are defined in Sect. B.2 and $\mathbf{F}_3$ and $\mathbf{P}_3$ are unknown functions to be determined. Substituting (B.31) into (B.12)

$$\begin{aligned}\frac{\mathrm{d}}{\mathrm{d}t}\left(\boldsymbol{\xi} + \varepsilon \mathbf{F}_1(t, \boldsymbol{\xi}) + \varepsilon^2 \mathbf{F}_2(t, \boldsymbol{\xi}) + \varepsilon^3 \mathbf{F}_3(t, \boldsymbol{\xi})\right) &= \varepsilon \mathbf{X}_1\left(t, \boldsymbol{\xi} + \varepsilon \mathbf{F}_1(t, \boldsymbol{\xi}) + \varepsilon^2 \mathbf{F}_2(t, \boldsymbol{\xi}) + \varepsilon^3 \mathbf{F}_3(t, \boldsymbol{\xi})\right) \\ &\quad + \varepsilon^2 \mathbf{X}_2\left(t, \boldsymbol{\xi} + \varepsilon \mathbf{F}_1(t, \boldsymbol{\xi}) + \varepsilon^2 \mathbf{F}_2(t, \boldsymbol{\xi}) + \varepsilon^3 \mathbf{F}_3(t, \boldsymbol{\xi})\right) \\ &\quad + \varepsilon^3 \mathbf{X}_3\left(t, \boldsymbol{\xi} + \varepsilon \mathbf{F}_1(t, \boldsymbol{\xi}) + \varepsilon^2 \mathbf{F}_2(t, \boldsymbol{\xi}) + \varepsilon^3 \mathbf{F}_3(t, \boldsymbol{\xi})\right).\end{aligned}$$

Expanding the RHS using Taylor series expansion

$$\begin{aligned}&\dot{\boldsymbol{\xi}} + \varepsilon \frac{\partial \mathbf{F}_1}{\partial t} + \varepsilon \nabla \mathbf{F}_1 \cdot \dot{\boldsymbol{\xi}} + \varepsilon^2 \frac{\partial \mathbf{F}_2}{\partial t} + \varepsilon^2 \nabla \mathbf{F}_2 \cdot \dot{\boldsymbol{\xi}} + \varepsilon^3 \frac{\partial \mathbf{F}_3}{\partial t} + \varepsilon^3 \nabla \mathbf{F}_3 \cdot \dot{\boldsymbol{\xi}} \\ &= \varepsilon \mathbf{X}_1 + \varepsilon^2 \nabla \mathbf{X}_1 \cdot \mathbf{F}_1 + \varepsilon^3 \nabla \mathbf{X}_1 \cdot \mathbf{F}_2 + \varepsilon^4 \nabla \mathbf{X}_1 \cdot \mathbf{F}_3 \\ &\quad + \tfrac{1}{2}\varepsilon^3 \begin{bmatrix} [\mathbf{F}_1 + \varepsilon \mathbf{F}_2 + \varepsilon^2 \mathbf{F}_3]^{\mathrm{T}} \nabla^2 X_{11} [\mathbf{F}_1 + \varepsilon \mathbf{F}_2 + \varepsilon^2 \mathbf{F}_3] \\ [\mathbf{F}_1 + \varepsilon \mathbf{F}_2 + \varepsilon^2 \mathbf{F}_3]^{\mathrm{T}} \nabla^2 X_{12} [\mathbf{F}_1 + \varepsilon \mathbf{F}_2 + \varepsilon^2 \mathbf{F}_3] \end{bmatrix} \\ &\quad + \varepsilon^2 \mathbf{X}_2 + \varepsilon^3 \nabla \mathbf{X}_2 \cdot \mathbf{F}_1 + \varepsilon^4 \nabla \mathbf{X}_2 \cdot \mathbf{F}_2 + \varepsilon^5 \nabla \mathbf{X}_2 \cdot \mathbf{F}_3 \\ &\quad + \tfrac{1}{2}\varepsilon^4 \begin{bmatrix} [\mathbf{F}_1 + \varepsilon \mathbf{F}_2 + \varepsilon^2 \mathbf{F}_3]^{\mathrm{T}} \nabla^2 X_{21} [\mathbf{F}_1 + \varepsilon \mathbf{F}_2 + \varepsilon^2 \mathbf{F}_3] \\ [\mathbf{F}_1 + \varepsilon \mathbf{F}_2 + \varepsilon^2 \mathbf{F}_3]^{\mathrm{T}} \nabla^2 X_{22} [\mathbf{F}_1 + \varepsilon \mathbf{F}_2 + \varepsilon^2 \mathbf{F}_3] \end{bmatrix} \\ &\quad + \varepsilon^3 \mathbf{X}_3 + \varepsilon^4 \nabla \mathbf{X}_3 \cdot \mathbf{F}_1 + \varepsilon^5 \nabla \mathbf{X}_3 \cdot \mathbf{F}_2 + \varepsilon^6 \nabla \mathbf{X}_3 \cdot \mathbf{F}_3.\end{aligned}$$

Substituting known functions $\mathbf{F}_1$, $\mathbf{F}_2$, $\mathbf{P}_1$, and $\mathbf{P}_2$ and neglecting higher-order terms yields

$$\frac{\partial \mathbf{F}_3}{\partial t} = \nabla\mathbf{X}_1 \cdot \mathbf{F}_2 - \nabla\mathbf{F}_2 \cdot \mathbf{P}_1 + \nabla\mathbf{X}_2 \cdot \mathbf{F}_1 - \nabla\mathbf{F}_1 \cdot \mathbf{P}_2 + \mathbf{X}_3 - \mathbf{P}_3 + \frac{1}{2}\begin{bmatrix} \mathbf{F}_1^{\mathrm{T}} \nabla^2 X_{11} \mathbf{F}_1 \\ \mathbf{F}_1^{\mathrm{T}} \nabla^2 X_{12} \mathbf{F}_1 \end{bmatrix}, \tag{B.33}$$

$\mathbf{P}_3$ is found to be

$$\mathbf{P}_3 = \frac{1}{T}\int_0^T \left\{ \nabla\mathbf{X}_1 \cdot \mathbf{F}_2 - \nabla\mathbf{F}_2 \cdot \mathbf{P}_1 + \nabla\mathbf{X}_2 \cdot \mathbf{F}_1 - \nabla\mathbf{F}_1 \cdot \mathbf{P}_2 + \mathbf{X}_3 + \frac{1}{2}\begin{bmatrix} \mathbf{F}_1^{\mathrm{T}} \nabla^2 X_{11} \mathbf{F}_1 \\ \mathbf{F}_1^{\mathrm{T}} \nabla^2 X_{12} \mathbf{F}_1 \end{bmatrix} \right\} dt. \tag{B.34}$$

It can be shown that given $\boldsymbol{\xi}(t)$ to be the solution of (B.32)

$$\left\| \mathbf{x}(t) - \boldsymbol{\xi}(t) - \varepsilon \mathbf{F}_1(t, \boldsymbol{\xi}(t)) - \varepsilon^2 \mathbf{F}_2(t, \boldsymbol{\xi}(t)) \right\| = O\left(\varepsilon^3\right), \tag{B.35}$$

for the time scale of $O(1/\varepsilon)$.

B.3 A Numerical Example

Here, a numerical example is presented which is taken from the test results of Chap. 9. The parameter values are selected according to the sample simulation results/measurements depicted in Fig. 9.21. The identified coupling stiffness, damping coefficient, and all the friction parameters are given, for reference, in Table B.1. The applied axial force and input angular velocity values are also listed in this table.

The results of the numerical averaging method are presented in Fig. B.1. In Fig. B.1b, the measurements are compared with simulation results showing the accuracy of modeling and the identified parameters similar to Fig. 9.21. In Fig. B.1a, the same simulation results are compared with the first-, second-, and third-order averaging. It can be seen that, for the selected parameter values, the first-order averaging has considerable error in predicting the steady-state amplitude of vibrations (a relative error of approximately 22%). The second-order averaging results, on the other hand, show significant improvement in both predicting the

Table B.1 Parameter values used in the higher-order averaging example

Parameter	Value	Friction parameter	Value
k	1.18 N m/rad	μ_1	0.218
c	19×10^{-5} Nms/rad	μ_2	0.0203
		μ_3	-4.47×10^{-4} s/rad
Inputs	Value	r_0	0.38 rad/s
Ω	35.6 rad/s	r_1	0.41 rad/s
R	153 N	s_μ	0.97

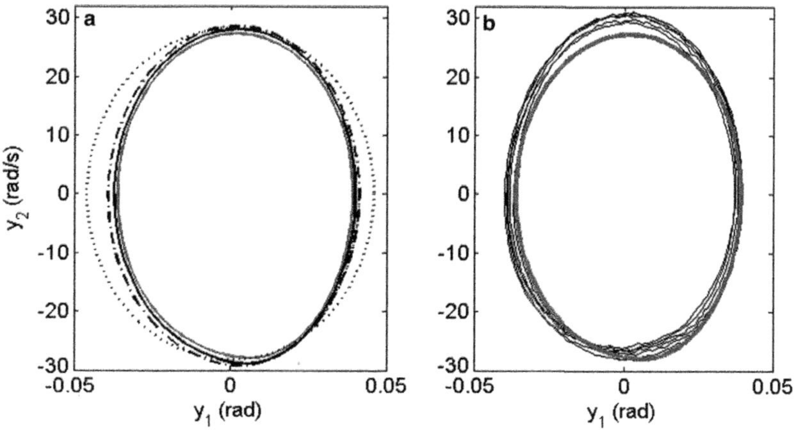

Fig. B.1 First-, second-, and third-order averaging results. (**a**) Numerical averaging results; *gray solid*: nonlinear system equation; *dotted black*: first-order averaging; *dashed-dot*: second-order averaging; *solid black*: third-order averaging, (**b**) *black*: measurements; *gray*: simulation results

steady-state vibration amplitude (relative error is approximately 5%) and conforming to the shape of the observed limit cycle. The accuracy of the approximation is further improved, though only slightly, by the third-order averaging which has approximately 4% relative error in predicting the amplitude of vibrations.

Appendix C: First-Order Averaging Applied to the 2-DOF Lead Screw Model with Axially Compliant Supports

In this appendix, the method of first-order averaging is used to analyze the 2-DOF model of Sect. 7.1.1. The equations of motion are given by (7.4). Neglecting F_0 and T_0 for simplicity and dividing the second equation of motion by $r_{\mathrm{m}}^2 \tan\lambda$, yield

$$\mathbf{M}(\mathbf{y},\dot{\mathbf{y}})\ddot{\mathbf{y}} + \mathbf{C}\dot{\mathbf{y}} + \mathbf{K}\mathbf{y} = \mathbf{f}(\mathbf{y},\dot{\mathbf{y}}), \tag{C.1}$$

where $\mathbf{y} = [y_1 \quad y_2]^{\mathrm{T}}$ defined by (7.1) and

$$\mathbf{M}(\mathbf{y},\dot{\mathbf{y}}) = \begin{bmatrix} I - \boldsymbol{\xi}(\mathbf{y},\dot{\mathbf{y}}) \tan\lambda m r_{\mathrm{m}}^2 & -\boldsymbol{\xi}(\mathbf{y},\dot{\mathbf{y}}) \tan\lambda m r_{\mathrm{m}}^2 \\ m & m + m_1 \end{bmatrix}, \tag{C.2}$$

$$\mathbf{C} = \begin{bmatrix} c & 0 \\ 0 & c_1 \end{bmatrix}, \tag{C.3}$$

$$\mathbf{K} = \begin{bmatrix} k & 0 \\ 0 & k_1 \end{bmatrix}, \tag{C.4}$$

and

$$\mathbf{f} = \begin{bmatrix} r_{\mathrm{m}}[\xi_0 - \xi(\mathbf{y}, \dot{\mathbf{y}})]R \\ 0 \end{bmatrix}. \tag{C.5}$$

Also

$$\xi(\mathbf{y}, \dot{\mathbf{y}}) = \frac{\mu_{\mathrm{s}}(\mathbf{y}, \dot{\mathbf{y}}) - \tan\lambda}{1 + \mu_{\mathrm{s}}(\mathbf{y}, \dot{\mathbf{y}}) \tan\lambda}, \tag{C.6}$$

and

$$\mu_{\mathrm{s}}(\mathbf{y}, \dot{\mathbf{y}}) = \mu(\dot{y}_1)\mathrm{sgn}(N(\mathbf{y}, \dot{\mathbf{y}}))\mathrm{sgn}(\dot{y}_1 + \Omega), \tag{C.7}$$

where the contact normal force is given by (5.32) and it is written as

$$N(\mathbf{y}, \dot{\mathbf{y}}) = \frac{\big((I/m) + (I/m_1) - r_{\mathrm{m}}^2 \tan\lambda\xi_0\big)R + r_{\mathrm{m}} \tan\lambda\big((Ik_1/m_1)y_2 + (Ic_1/m_1)\dot{y}_2 + ky_1 + c\dot{y}_1\big)}{(\cos\lambda + \mu_{\mathrm{s}}(\mathbf{y}, \dot{\mathbf{y}}) \sin\lambda)\big((I/m) + (I/m_1) - r_{\mathrm{m}}^2 \tan\lambda\xi(\mathbf{y}, \dot{\mathbf{y}})\big)}. \tag{C.8}$$

Assuming $\mathbf{M}$ to be nonsingular (and dropping the arguments for brevity)

$$\mathbf{M}^{-1} = \frac{1}{1 - r_{\mathrm{m}}^2 \tan\lambda\xi(mm_1/I(m+m_1))} \begin{bmatrix} \frac{1}{I} & \dfrac{\xi m r_{\mathrm{m}}^2 \tan\lambda}{I(m+m_1)} \\ -\dfrac{m}{I(m+m_1)} & \dfrac{I - \xi m r_{\mathrm{m}}^2 \tan\lambda}{I(m+m_1)} \end{bmatrix}. \tag{C.9}$$

Multiplying both sides of (C.1) by $\mathbf{M}^{-1}$, one may find

$$\left(1 + r_{\mathrm{m}}^2 \tan\lambda\xi \frac{mm_1}{I(m+m_1)}\right)\ddot{\mathbf{y}} + \hat{\mathbf{C}}\dot{\mathbf{y}} + \hat{\mathbf{K}}\mathbf{y} = \hat{\mathbf{f}}, \tag{C.10}$$

where

$$\hat{\mathbf{C}} = \begin{bmatrix} \dfrac{c}{I} & \dfrac{c_1 \xi m r_{\mathrm{m}}^2 \tan\lambda}{I(m+m_1)} \\ -\dfrac{cm}{I(m+m_1)} & \dfrac{c_1\big(I - \xi m r_{\mathrm{m}}^2 \tan\lambda\big)}{I(m+m_1)} \end{bmatrix}, \tag{C.11}$$

$$\hat{\mathbf{K}} = \begin{bmatrix} \dfrac{k}{I} & \dfrac{k_1 \xi m r_{\mathrm{m}}^2 \tan\lambda}{I(m+m_1)} \\ -\dfrac{km}{I(m+m_1)} & \dfrac{k_1\left(I - \xi m r_{\mathrm{m}}^2 \tan\lambda\right)}{I(m+m_1)} \end{bmatrix}, \tag{C.12}$$

$$\hat{\mathbf{f}} = \begin{bmatrix} 1 \\ \frac{m}{(m+m_1)} \end{bmatrix} \frac{R}{I} r_{\mathrm{m}} (\xi - \xi_0), \tag{C.13}$$

The two natural frequencies of the undamped unperturbed system are

$$\omega_1 = \sqrt{\frac{k}{I}}, \quad \omega_2 = \sqrt{\frac{k_1}{m+m_1}}. \tag{C.14}$$

Let

$$\widehat{m} = r_{\mathrm{m}}^2 \tan\lambda \frac{m}{I}, \tag{C.15}$$

$$\widehat{\mu}_{\mathrm{s}} = \frac{\mu_{\mathrm{s}}}{\tan\lambda}. \tag{C.16}$$

Taking $\varepsilon \equiv \tan\lambda$ as the small parameter, it is assumed that $\widehat{m}$ and $\widehat{\mu}_{\mathrm{s}}$ are $O(1)$ with respect to ε. We can write the following asymptotic expansion for ξ

$$\xi = \varepsilon r_{\mathrm{m}}\left(\widehat{\mu}_{\mathrm{s}} - 1\right) - \varepsilon^3 r_{\mathrm{m}} \widehat{\mu}_{\mathrm{s}} \left(\widehat{\mu}_{\mathrm{s}} - 1\right) + \cdots. \tag{C.17}$$

Also

$$\left(1 - \varepsilon \frac{\widehat{m} m_1}{m+m_1} \frac{\widehat{\mu}_{\mathrm{s}} - 1}{1 + \varepsilon^2 \widehat{\mu}_{\mathrm{s}}}\right)^{-1} = 1 + \varepsilon \frac{\widehat{m} m_1}{m+m_1}\left(\widehat{\mu}_{\mathrm{s}} - 1\right) + O\left(\varepsilon^3\right). \tag{C.18}$$

Define the nondimensionalized time as $\tau = \omega_1 t$. Derivative with respect to τ is given by

$$\frac{\mathrm{d}(\cdot)}{\mathrm{d}t} = \omega_1 \frac{\mathrm{d}(\cdot)}{\mathrm{d}\tau}, \quad (\cdot)' \equiv \frac{\mathrm{d}(\cdot)}{\mathrm{d}\tau}. \tag{C.19}$$

Also, limiting the analysis to weakly damped systems, we take

$$\varepsilon \widehat{c} = \frac{c}{I\omega_1}, \quad \varepsilon \widehat{c}_1 = \frac{c_1}{\omega_1 (m+m_1)}. \tag{C.20}$$

Assume $|\Omega|/\omega_1 = \rho\varepsilon$ where ρ is $O(1)$ with respect to ε. Let,

$$\mathbf{y} = \varepsilon\rho\mathbf{z}. \tag{C.21}$$

Substituting (C.14)–(C.16), (C.20), and (C.21) into (C.22) and using (C.17)–(C.19) and truncating $O(\varepsilon^2)$ terms yields

$$\mathbf{z}'' + \tilde{\mathbf{K}}_0\mathbf{z} = -\varepsilon\tilde{\mathbf{C}}_1\mathbf{z}' + \varepsilon\widehat{m}(1-\alpha)\left(1-\widehat{\mu}_\mathrm{s}\right)\tilde{\mathbf{K}}_0\mathbf{z} - \varepsilon\tilde{\mathbf{K}}_1\mathbf{z} + \varepsilon\tilde{\mathbf{f}}_0, \tag{C.22}$$

where

$$\alpha = \frac{m}{m+m_1}, \tag{C.23}$$

$$\tilde{\mathbf{K}}_0 = \begin{bmatrix} 1 & 0 \\ -\alpha & \eta^2 \end{bmatrix}, \tag{C.24}$$

$$\tilde{\mathbf{K}}_1 = \eta^2\left(1-\widehat{\mu}_\mathrm{s}\right)\widehat{m}\begin{bmatrix} 0 & -1 \\ 0 & 1 \end{bmatrix}, \tag{C.25}$$

$$\tilde{\mathbf{C}}_1 = \begin{bmatrix} \widehat{c} & 0 \\ -\alpha\widehat{c} & \widehat{c}_1 \end{bmatrix}, \tag{C.26}$$

$$\tilde{\mathbf{f}}_0 = \widehat{R}\left(\widehat{\mu}_0 \ \mathrm{sgn}\ (R\Omega) - \widehat{\mu}_\mathrm{s}\right)\begin{bmatrix} 1 \\ -\alpha \end{bmatrix}, \tag{C.27}$$

where

$$\widehat{R} = \frac{\omega_1}{|\Omega|}\frac{r_\mathrm{m}R}{k}, \tag{C.28}$$

is considered to be $O(1)$ with respect to ε. Also, η is the ratio of the two natural frequencies

$$\eta = \frac{\omega_2}{\omega_1}. \tag{C.29}$$

Assuming that the two natural frequencies of the unperturbed system are widely apart, let

$$\mathbf{T} = \begin{bmatrix} 1 & 0 \\ \alpha\frac{1}{\eta^2-1} & 1 \end{bmatrix}, \tag{C.30}$$

then

$$\boldsymbol{\omega}^2 = \mathbf{T}^{-1}\hat{\mathbf{K}}_0\mathbf{T} = \begin{bmatrix} 1 & 0 \\ 0 & \eta^2 \end{bmatrix}. \tag{C.31}$$

Let $\mathbf{z} = \mathbf{T}\mathbf{w}$, systems' equations (C.22) are transformed to

$$\mathbf{w}'' + \boldsymbol{\omega}^2\mathbf{w} = -\varepsilon\mathbf{T}^{-1}\tilde{\mathbf{C}}_1\mathbf{T}\mathbf{w}' + \varepsilon\hat{m}(1-\alpha)\left(1-\hat{\mu}_s\right)\boldsymbol{\omega}^2\mathbf{w} - \varepsilon\mathbf{T}^{-1}\tilde{\mathbf{K}}_1\mathbf{T}\mathbf{w} + \varepsilon\mathbf{T}^{-1}\tilde{\mathbf{f}}_0. \tag{C.32}$$

Finally, let[1]

$$\mathbf{w} = \begin{bmatrix} v_1\cos\tau - v_2\sin\tau \\ v_3\cos\eta\tau - \eta^{-1}v_4\sin\eta\tau \end{bmatrix}, \quad \mathbf{w}' = \begin{bmatrix} -v_1\sin\tau - v_2\cos\tau \\ -\eta v_3\sin\eta\tau - v_4\cos\eta\tau \end{bmatrix}. \tag{C.33}$$

Substituting (C.33) into (C.32) and expanding yield the following four first-order differential equations in standard form:

$$\begin{aligned} v'_1 = &-\varepsilon\hat{c}(v_1\sin\tau + v_2\cos\tau)\sin\tau \\ &-\varepsilon\hat{m}\left(1+\frac{\alpha}{\eta^2-1}\right)\left(1-\hat{\mu}_s\right)(v_1\cos\tau - v_2\sin\tau)\sin\tau \\ &-\varepsilon\hat{m}\eta^2\left(1-\hat{\mu}_s\right)\left(v_3\cos\eta\tau - \eta^{-1}v_4\sin\eta\tau\right)\sin\tau \\ &-\varepsilon\hat{R}\left(\hat{\mu}_0\,\mathrm{sgn}\,(R\Omega) - \hat{\mu}_s\right)\sin\tau, \end{aligned} \tag{C.34}$$

$$\begin{aligned} v'_2 = &-\varepsilon\hat{c}(v_1\sin\tau + v_2\cos\tau)\cos\tau \\ &-\varepsilon\hat{m}\left(1+\frac{\alpha}{\eta^2-1}\right)\left(1-\hat{\mu}_s\right)(v_1\cos\tau - v_2\sin\tau)\cos\tau \\ &-\varepsilon\hat{m}\eta^2\left(1-\hat{\mu}_s\right)\left(v_3\cos\eta\tau - \eta^{-1}v_4\sin\eta\tau\right)\cos\tau \\ &-\varepsilon\hat{R}\left(\hat{\mu}_0\,\mathrm{sgn}\,(R\Omega) - \hat{\mu}_s\right)\cos\tau, \end{aligned} \tag{C.35}$$

$$\begin{aligned} v'_3 = &-\varepsilon\frac{\alpha(\hat{c}_1 - \eta^2\hat{c})}{\eta(\eta^2-1)}(v_1\sin\tau + v_2\cos\tau)\sin\eta\tau - \varepsilon\frac{\hat{c}_1}{\eta}(\eta v_3\sin\eta\tau + v_4\cos\eta\tau)\sin\eta\tau \\ &-\varepsilon\eta\hat{m}(1-\alpha)\left(1-\hat{\mu}_s\right)\left(v_3\cos\eta\tau - \eta^{-1}v_4\sin\eta\tau\right)\sin\eta\tau \\ &+\varepsilon\eta\left(1-\hat{\mu}_s\right)\hat{m}\alpha\frac{1}{\eta^2-1}\left(\alpha\frac{1}{\eta^2-1}+1\right)(v_1\cos\tau - v_2\sin\tau)\sin\eta\tau \\ &+\varepsilon\eta\left(1-\hat{\mu}_s\right)\hat{m}\left(\alpha\frac{1}{\eta^2-1}+1\right)\left(v_3\cos\eta\tau - \eta^{-1}v_4\sin\eta\tau\right)\sin\eta\tau \\ &+\varepsilon\alpha\frac{\eta}{\eta^2-1}\hat{R}\left(\hat{\mu}_0\,\mathrm{sgn}\,(R\Omega) - \hat{\mu}_s\right)\sin\eta\tau \end{aligned} \tag{C.36}$$

---

[1]Note that the more convenient choice of "amplitude/phase" transform (see Sect. 5.3.2) is not used here since the resulting differential equations would have been singular at the origin.

$$
\begin{aligned}
v'_4 = & -\varepsilon \frac{\alpha\left(\widehat{c}_1 - \eta^2 \widehat{c}\right)}{(\eta^2 - 1)} (v_1 \sin\tau + v_2 \cos\tau) \sin\eta\tau - \varepsilon\widehat{c}_1 (\eta v_3 \sin\eta\tau + v_4 \cos\eta\tau) \cos\eta\tau \\
& - \varepsilon\eta\widehat{m}(1-\alpha)\left(1 - \widehat{\mu}_s\right)(\eta v_3 \cos\eta\tau - v_4 \sin\eta\tau) \cos\eta\tau \\
& + \varepsilon\eta^2\left(1 - \widehat{\mu}_s\right)\widehat{m}\alpha \frac{1}{\eta^2 - 1}\left(\alpha\frac{1}{\eta^2 - 1} + 1\right)(v_1 \cos\tau - v_2 \sin\tau)\cos\eta\tau \\
& + \varepsilon\eta\left(1 - \widehat{\mu}_s\right)\widehat{m}\left(\alpha\frac{1}{\eta^2 - 1} + 1\right)(\eta v_3 \cos\eta\tau - v_4 \sin\eta\tau)\cos\eta\tau \\
& + \varepsilon\alpha\frac{\eta^2}{\eta^2 - 1}\widehat{R}\left(\widehat{\mu}_0 \operatorname{sgn}(R\Omega) - \widehat{\mu}_s\right)\cos\eta\tau,
\end{aligned}
\tag{C.37}
$$

where $O(\varepsilon^2)$ terms are neglected. From this point on, we assume that $N(\mathbf{v}, \mathbf{v}') > 0$. Also note that the argument of $\widehat{\mu}_s(\dot{\theta})$ (lead screw angular velocity) is given by

$$
\dot{\theta} = \Omega - |\Omega|(v_1 \sin\tau + v_2 \cos\tau). \tag{C.38}
$$

It is more convenient to express the above system of first-order differential equations as

$$
\begin{aligned}
\mathbf{v}' &= \varepsilon\mathbf{g}(\mathbf{v}, \boldsymbol{\psi}, \varepsilon), \\
\psi'_1 &= 1, \\
\psi'_2 &= \eta,
\end{aligned}
\tag{C.39}
$$

where $\mathbf{v} = \begin{bmatrix} v_1 & v_2 & v_3 & v_4 \end{bmatrix}^{\mathrm{T}}$. The right-hand side of (C.39) is quasiperiodic in τ (i.e., it is 2π periodic with respect to both ψ_1 and ψ_2). Hence, the first-order averaged equations can be derived from [56]

$$
\bar{\mathbf{v}}' = \frac{\varepsilon}{4\pi^2}\int_0^{2\pi}\int_0^{2\pi} \mathbf{g}(\bar{\mathbf{v}}, \boldsymbol{\psi}, 0)\,\mathrm{d}\psi_1\,\mathrm{d}\psi_2. \tag{C.40}
$$

After some simplifications, carrying out the integration (C.34)–(C.37) yields (dropping the bars)

$$
\begin{aligned}
v'_1 = & -\varepsilon\frac{\widehat{c}}{2}v_1 + \varepsilon\widehat{R}\frac{1}{2\pi}\int_0^{2\pi}\widehat{\mu}_s \sin\psi_1\,\mathrm{d}\psi_1 - \varepsilon\widehat{m}\left(1 + \frac{\alpha}{\eta^2 - 1}\right)\frac{1}{2\pi} \\
& \times \int_0^{2\pi}\left(1 - \widehat{\mu}_s\right)\left(v_1 \sin\psi_1 \cos\psi_1 - \frac{v_2}{\omega_1}\sin^2\psi_1\right)\mathrm{d}\psi_1,
\end{aligned}
\tag{C.41}
$$

$$v_2' = -\varepsilon\frac{\widehat{c}}{2}v_2 + \varepsilon\widehat{R}\frac{1}{2\pi}\int_0^{2\pi}\widehat{\mu}_s\cos\psi_1 \mathrm{d}\psi_1 - \varepsilon\widehat{m}\left(1+\frac{\alpha}{\eta^2-1}\right)\frac{1}{2\pi}$$
$$\times\int_0^{2\pi}\left(1-\widehat{\mu}_s\right)\left(v_1\cos^2\psi_1 - \frac{v_2}{\omega_1}\sin\psi_1\cos\psi_1\right)\mathrm{d}\psi_1, \qquad \text{(C.42)}$$

$$v_3' = -\varepsilon\frac{\widehat{c}_1}{2}v_3 + \varepsilon\widehat{m}(1-\alpha)v_4\frac{1}{4\pi}\int_0^{2\pi}\left(1-\widehat{\mu}_s\right)\mathrm{d}\psi_1$$
$$-\varepsilon\widehat{m}\left(\alpha\frac{1}{\eta^2-1}+1\right)v_4\frac{1}{4\pi}\int_0^{2\pi}\left(1-\widehat{\mu}_s\right)\mathrm{d}\psi_1, \qquad \text{(C.43)}$$

$$v_4' = -\varepsilon\frac{\widehat{c}_1}{2}v_4 - \varepsilon\eta^2\widehat{m}(1-\alpha)v_3\frac{1}{4\pi}\int_0^{2\pi}\left(1-\widehat{\mu}_s\right)\mathrm{d}\psi_1$$
$$+\varepsilon\eta^2\widehat{m}\left(\alpha\frac{1}{\eta^2-1}+1\right)v_3\frac{1}{4\pi}\int_0^{2\pi}\left(1-\widehat{\mu}_s\right)\mathrm{d}\psi_1, \qquad \text{(C.44)}$$

where $\widehat{\mu}_s = \widehat{\mu}_s(\Omega - |\Omega|(v_1\sin\psi_1 + v_2\cos\psi_1))$. Introducing the polar coordinates

$$\begin{aligned} v_1 &= a_1\cos\beta_1, \\ v_2 &= a_1\sin\beta_1, \\ v_3 &= a_2\cos\beta_2, \\ v_4 &= a_2\eta\sin\beta_2, \end{aligned} \qquad \text{(C.45)}$$

the amplitude equations are found as

$$a_1' = -\varepsilon\frac{\widehat{c}}{2}a_1 + \varepsilon\frac{\widehat{R}}{2\pi}\int_0^{2\pi}\widehat{\mu}_s\sin(\psi_1+\beta_1)\mathrm{d}\psi_1 - \varepsilon\widehat{m}\left(1+\frac{\alpha}{\eta^2-1}\right)\frac{a_1}{2\pi}$$
$$\times\int_0^{2\pi}\left(1-\widehat{\mu}_s\right)\cos(\psi_1+\beta_1)\sin(\psi_1+\beta_1)\mathrm{d}\psi_1, \qquad \text{(C.46)}$$

$$a_2' = -\varepsilon\frac{\widehat{c}_1}{2}a_2. \qquad \text{(C.47)}$$

This first equation is similar to the 1-DOF case studied in Sect. 6.3.3. In fact, if the same limitations are considered as in Sect. 6.3.1 (i.e., $\mathrm{sgn}(\dot{\theta}) = \mathrm{sgn}(N) = 1$), it further simplifies to

$$a_1' = -\varepsilon\frac{\widehat{c}}{2}a_1 + \varepsilon\frac{\widehat{R}}{2\pi}\int_0^{2\pi}\widehat{\mu}\sin\psi_1\,\mathrm{d}\psi_1, \qquad \text{(C.48)}$$

where $\widehat{\mu} = \widehat{\mu}(\Omega(1 - a_1 \sin \psi_1))$ and is defined similar to (6.36) by

$$\widehat{\mu} = \left(\widehat{\mu}_1 + \widehat{\mu}_2 e^{r_0 \Omega a_1 \sin \psi_1}\right) \left(1 - r_2 e^{r_1 \Omega a_1 \sin \psi_1}\right) - \widehat{\mu}_3 a_1 \sin \psi_1, \tag{C.49}$$

and $\widehat{\mu}_1$, $\widehat{\mu}_2$, and $\widehat{\mu}_3$ are given by (6.37), (6.38), and (6.39), respectively.

Notice that (C.48) is exactly the same as (6.35) if a_1 is replaced by a. The first-order averaged equations for a_1, (C.48), and a_2, (C.47), are decoupled. Furthermore, (C.47) shows that, to this order of approximations, the vibration component with the frequency ω_2 dies out exponentially, independent of a_1.

References

1. Gallina P, Giovagnoni M (2002) Design of screw jack mechanism to avoid self-excited vibrations. J Dyn Syst Meas Control 124:477–480
2. Hollander KW, Sugar TG (2006) Design of lightweight lead screw actuators for wearable robotic applications. Trans ASME J Mech Des 128:644–648
3. Buck R (2006) A few pointers for selecting pumps. Med Des (November). http://medicaldesign.com/pumps-valves/few_pointers_selecting/
4. Ibrahim RA (1994) Friction-induced vibration, chatter, squeal, and chaos. Part I: mechanics of contact and friction. Appl Mech Rev 47(7):209–226
5. Awrejcewicz J, Olejnik P (2005) Analysis of dynamic systems with various friction laws. Appl Mech Rev 58:389–411
6. Ibrahim RA (1994) Friction-induced vibration, chatter, squeal, and chaos. Part II: dynamics and modeling. Appl Mech Rev 47(7):227–253
7. Kinkaid NM, O'Reilly OM, Papadopoulos P (2003) Review: automotive disc brake squeal. J Sound Vib 267:105–166
8. D'Souza AF, Dweib AH (1990) Self-excited vibrations induced by dry friction, part 2: stability and limit-cycle analysis. J Sound Vib 137(2):177–190
9. Popp K (2005) Modelling and control of friction-induced vibrations. Math Comput Model Dyn Syst 11(3):345–369
10. Feeny B, Guran A, Hinrichs N, Popp K (1998) Historical review on dry friction and stick–slip phenomena. Appl Mech Rev 51(5):321–341
11. Armstrong-Helouvry B, Dupont P, Canudas de Wit C (1994) A survey of models, analysis tools and compensation methods for the control of machines with friction. Automatica 30(7):1083–1138
12. Berger EJ (2002) Friction modeling for dynamic system simulation. Appl Mech Rev 55(6):535–577
13. Canudas de Wit C, Olsson H, Åström KJ, Lischinsky P (1995) A new model for control of systems with friction. IEEE Trans Autom Control 40(3):419–425
14. Armstrong-Helouvry B (1990) Stick–slip arising from stribeck friction. In: Proceedings of the IEEE international conference on robotics and automation, Cincinnati, OH, pp 1377–1382
15. Armstrong-Helouvry B (1991) Control machines with friction. Kluwer, Boston
16. Marton L, Lantos B (2007) Modeling, identification, and compensation of stick–slip friction. IEEE Trans Ind Electron 54(1):511–521
17. Leine RI, van Campen DH, de Kraker A (1998) Stick–slip vibrations induced by alternate friction models. Nonlinear Dyn 16:41–54
18. Olofsson U, Ekerfors L (1991) Friction-induced noise in screw–nut transmissions. Wear 148:25–37

19. Dupont PE (1992) The effect of coulomb friction on the existence and uniqueness of the dynamics problem. In: Proceedings of the 1992 IEEE international conference on robotics and automation, Nice, France, pp 1442–1447
20. Gallina P (2005) Vibration in screw jack mechanisms: experimental results. J Sound Vib 282:1025–1041
21. Oledzki AA (1995) Modeling and simulation of self-locking drives. Mech Mach Theory 30(6):929–942
22. Scott P (2004) Ball-screw basics: debunking the myths. Mach Des 76(13):88–92
23. Otsuka J (1992) Nanometer level positioning using three kinds of lead screws. Nanotechnology 3:29–36
24. Shimokohbe A (1996) The state of the art of nano-meter motion control in Japan. In: Proceedings of the international conference on trends in industrial measurements automation, pp 2–8
25. Shimokohbe A (1999) The development of nano-meter positioning technology in Japan. In: Proceedings of the international conference on advanced manufacturing technology, pp 83–87
26. Awabdy BA, Shih W-C, Auslander DM (1998) Nanometer positioning of a linear motion stage under static loads. IEEE ASME Trans Mechatron 3(2):113–119
27. Erkorkmaz K, Altintas Y (2001) High speed CNC system design. Part II: modeling and identification of feed drives. Int J Mach Tools Manufact 41:1487–1509
28. Varanasi KK, Nayfeh SA (2004) The dynamics of lead-screw drives: low-order modeling and experiments. J Dyn Syst Meas Control 126:388–396
29. Chen CL, Jang MJ, Lin KC (2004) Modeling and high-precision control of a ball-screw-driven stage. Precis Eng 28:483–495
30. Whalley R, Ebrahimi M, Abdul-Ameer AA (2005) Hybrid modelling of machine tool axis drives. Int J Mach Tools Manufact 45:1560–1576
31. Sato K, Fujimoto H, Osada H, Shimokohbe A (1995) Dynamic behaviour of lead screw mechanism with backlash. Jpn Soc Mech Eng C 61(584):1423–1430
32. Sato K, Murayama Y, Imada S, Shimokohbe A (1995) Control and elimination of lead screw backlash for ultra-precision positioning. JSME Int J Ser C 38(1):36–41
33. MacKenzie RV (1961) Screw threads – design, selection, and specification. Industrial, New York
34. Shigley JE, Mischke CR (1989) Mechanical engineering design, 5th edn. McGraw-Hill, New York
35. Spotts MF (1971) Design of machine elements, 4th edn. Prentice-Hall, Englewood Cliffs
36. —— High precision ballscrew and leadscrew driven tables, parker products catalog 8092/USA Section B. http://www.parker.com/literature/Literature%20Files/emn/SectionB.pdf. Accessed 30 Nov 2008
37. —— Miniature precision stages, parker products catalog 8092/USA Section C. http://www.parker.com/literature/Literature%20Files/emn/SectionC.pdf. Accessed 30 Nov 2008
38. —— (2008) Ball screws vs. lead screws. Motion Syst Des, Jan. 2008. http://motionsystemdesign.com/mag/ballscrews_vs_lead/ [Accessed 30 Nov 2008]
39. Solon T (2006) Taking the lead. Motion Syst Des 48(1):20–26
40. —— (2003) Lead screws roll over ball screws. Mach Des 75(23):94–98
41. —— (2008) Lead and ball screw catalog. Thomson BSA. http://www.thomsonbsa.com/pdf/ThomsonBSA-Lead-and-Ball-Screw-Catalog.pdf. Accessed 30 Nov 2008
42. —— (2008) Precision rolled lead screws – metric series. Thomson BSA. http://www.thomsonbsa.com/pdf/ThomsonBSA-Precision-Rolled-Lead-Screws-Metric-Series.pdf. Accessed 30 Nov 2008
43. —— (2008) PowerAc precision lead screw assemblies. Nook industries. http://www.nookindustries.com/pdf/NookAcmeScrew.pdf. Accessed 30 Nov 2008
44. —— (2008) Kerk motion products catalog. Kerk Motion Products, Inc. http://www.kerkmotion.com/pdf/kerk-catalog-2008.pdf. Accessed 30 Nov 2008

45. — (2008) Which screw? Picking the right technology. Tolomatic, Inc. http://www.mail.tolomatic.com/archives/pdfs/9900–4644_00_WhichScrew.pdf. Accessed 30 Nov 2008
46. — (2008) Lead screw nut: general catalog – (A) technical descriptions of products. THK. http://www.thk.com/documents/us_pdf/products/generalA/en_A_1_general.pdf. Accessed 30 Nov 2008
47. Stolarski TA (1990) Tribology in machine design. Butterworth-Heinemann, London
48. Huseyin K (1978) Vibrations and stability of multiple parameter systems. Noordhoff, Amsterdam
49. Ziegler H (1968) Principles of structural stability. Blaisdell, Waltham
50. Inman DJ (2006) Vibration with control. Wiley, London
51. Anh LX (2003) Dynamics of mechanical systems with coulomb friction. Springer, New York
52. Leine RI, Nijmeijer H (2004) Dynamics and bifurcations of non-smooth mechanical systems. Lecture notes in applied and computational mechanics, vol. 18. Springer, Heidelberg
53. Leine RI, Brogliato B, Nijmeijer H (2002) Periodic motion and bifurcations induced by the Painlevé paradox. Eur J Mech A Solids 21(5):869–896
54. Chen C-T (1999) Linear system theory and design, 3rd edn. Oxford University Press, Oxford
55. Bogoliubov NN, Mitropolsky YA (1961) Asymptotic methods in the theory of non-linear oscillations, 2nd edn. (translated from Russian). Hindustan, Delhi
56. Sanders J, Verhulst F, Murdock J (2007) Averaging methods in nonlinear dynamical systems, 2nd edn. Springer, New York
57. Fidlin A (2006) Nonlinear oscillations in mechanical engineering. Springer, New York
58. Khalil HK (2001) Nonlinear systems, 3rd edn. Prentice-Hall, Englewood Cliffs
59. Feeny BF (1996) The nonlinear dynamics of oscillators with stick–slip friction. In: Guran A, Pfeiffer F, Popp K (eds) Dynamics with friction. World Scientific, River Edge, pp 36–92
60. Popp K, Rudolph M (2004) Vibration control to avoid stick–slip motion. J Vib Control 10(11):1585–1600
61. Hetzler H, Schwarzer D, Seemann W (2007) Analytical investigation of steady-state stability and Hopf-bifurcations occurring in sliding friction oscillators with application to low-frequency disc brake noise. Commun Nonlinear Sci Numer Simul 12:83–99
62. Thomsen JJ, Fidlin A (2003) Analytical approximations for stick–slip vibration amplitudes. Int J Non-Linear Mech 38:389–403
63. Oden JT, Martins JAC (1985) Models and computational methods for dynamic friction phenomena. Comput Meth Appl Mech Eng 52:527–634
64. Chatterjee S (2007) Non-linear control of friction-induced self-excited vibration. Int J Non Lin Mech 42:459–469
65. Majd VJ, Simaan MA (1995) A continuous friction model for servo systems with stiction. In: Proceedings of the IEEE conference on control applications, pp. 296–301
66. Beloiu DM, Ibrahim RA (2006) Analytical and experimental investigations of disc brake noise using the frequency-time domain. Struct Control Health Monit 13:277–300
67. van de Vrande BL, van Campen DH, de Kraker A (1999) An approximate analysis of dry-friction-induced stick–slip vibrations by a smoothing procedure. J Nonlinear Dyn 19(2):159–171
68. Henrici P (1974) Applied and computational complex analysis, vol. I. Wiley, New York
69. Nayfeh AH, Balachandran B (1995) Applied nonlinear dynamics. Wiley, New York
70. Brockley CA, Ko PL (1988) Investigation of rail corrugation using friction-induced vibration theory. Wear 128(1):99–106
71. Elmer FJ (1997) Nonlinear dynamics of dry friction. J Phys A Math Gen 30:6057–6063
72. Sundararajan C (1975) The vibration and stability of elastic systems subjected to follower forces. Shock Vib Digest 7(6):89–105
73. Sethna PR, Schapiro SM (1977) Nonlinear behaviour of flutter unstable dynamical systems with gyroscopic and circulatory forces. J Appl Mech 44:755–762
74. Ono K, Chen JS, Bogy DB (1991) Stability analysis of the head-disk interface in a flexible disc drive. ASME J Appl Mech 58:1005–1014

75. Mottershead JE, Chan SN (1995) Flutter instability of circular discs with frictional followers loads. J Vib Acoust 117:161–163
76. Hoffmann N, Gaul L (2003) Effects of damping on mode-coupling instability in friction induced oscillations. ZAMM 83(8):524–534
77. Hoffmann N, Bieser S, Gaul L (2004) Harmonic balance and averaging techniques for stick–slip limit-cycle determination in mode-coupling friction self-excited systems. Tech Mech 24(3–4):185–197
78. Sinou J-J, Fritz G, Jézéquel L (2007) The role of damping and definition of the robust damping factor for a self-exciting mechanism with constant friction. J Vib Acoust 129: 297–306
79. Hoffmann N, Fischer M, Allgaier R, Gaul L (2002) A minimal model for studying properties of the mode-coupling type instability in friction induced oscillations. Mech Res Commun 29:197–205
80. Hoffmann N (2006) Transient growth and stick–slip in sliding friction. J Appl Mech 73(4): 642–647
81. Hoffmann NP (2007) Linear stability of steady sliding in point contacts with velocity dependent and LuGre type friction. J Sound Vib 301:1023–1034
82. Fritz G, Sinou J-J, Duffala J-M, Jézéquel L (2007) Investigation of the relationship between damping and mode-coupling patterns in case of brake squeal. J Sound Vib 307:591–609
83. Sinou J-J, Jézéquel L (2007) Mode coupling instability in friction-induced vibrations and its dependency on system parameters including damping. Eur J Mech A Solids 26:106–122
84. Sinou J-J, Jézéquel L (2007) The influence of damping on the limit cycles for a self-exciting mechanism. J Sound Vib 304(3–5):875–893
85. Fritz G, Sinou J-J, Duffal J-M, Jézéquel L (2007) Effects of damping on brake squeal coalescence patterns – application on a finite element model. Mech Res Commun 34: 181–190
86. Sinou J-J, Fritz G, Jézéquel L (2007) The role of damping and definition of the robust damping factor for a self-exciting mechanism with constant friction. J Vib Acoust 129(3): 297–306
87. Hervé B, Sinou J-J, Mahéa H, Jézéquel L (2008) Analysis of squeal noise and mode coupling instabilities including damping and gyroscopic effects. Eur J Mech Solid 27:141–160
88. Sinou J-J, Jézéquel L (2008) On the stabilizing and destabilizing effects of damping in a non-conservative pin-disc system. Acta Mech 199:43–52
89. Chevillot F, Sinou J-J, Mazet G-B, Hardouin N, Jézéquel L (2008) The destabilization paradox applied to friction-induced vibrations in an aircraft braking system. Arch Appl Mech 78:949–963
90. Rusli M, Okuma M (2007) Effect of surface topography on mode-coupling model of dry contact sliding systems. J Sound Vib 308:721–734
91. von Wagner U, Hochlenert D, Hagedorn P (2007) Minimal models for disk brake squeal. J Sound Vib 302:527–539
92. Huang J, Krousgrill CM, Bajaj AK (2007) An efficient approach to estimate critical value of friction coefficient in brake squeal analysis. J Appl Mech 74(3):534–541
93. Chatterjee S (2008) On the design criteria of dynamic vibration absorbers for controlling friction-induced oscillations. J Vib Control 14(3):397–415
94. Kang J (2008) Parametric study on friction-induced coupled oscillator. Proc IME C J Mech Eng Sci 222(8):1381–1387
95. Painlevé P (1895) Sur les lois du frottement de glissement. C R Séances Acad Sci 121:112–115
96. Brogliato B (1999) Nonsmooth mechanics models, dynamics and control, 2nd edn. Communications and control engineering. Springer, London
97. Dupont PE, Yamajako SP (1994) Jamming and wedging in constrained rigid-body dynamics. In: Proceedings of the 1994 IEEE international conference on robotics and automation, vol 3, pp 2349–2354

98. Hoffmann N, Gaul L (2004) A sufficient criterion for the onset of sprag-slip oscillations. Arch Appl Mech 73:650–660
99. Dupont PE (1992) The use of compliance to resolve the existence and uniqueness of the forward dynamics solution with coulomb friction. In: Proceedings of CSME forum 1992, Montreal, Canada
100. Dupont PE, Yamajako SP (1997) Stability of frictional contact in constrained rigid-body dynamics. IEEE Trans Robot 13(2):230–236
101. Deryabin MV, Kozlov VV (1995) A theory of systems with unilateral constraints. J Appl Math Mech 59(4):505–512
102. Zhao Z, Chen B, Liu C, Jin H (2004) Impact model resolution on painleve's paradox. Acta Mech Sin 20(6):659–660
103. Samsonov VA (2005) The dynamics of a brake shoe and impact generated by friction. J Appl Math Mech 69:816–824
104. Glocker C (2001) Set-valued force laws, dynamics of non-smooth systems. Lecture Notes in Applied Mechanics, vol. 1. Springer, Berlin
105. Pfeiffer F, Glocker C (1996) Multibody dynamics with unilateral contacts. Wiley, New York
106. Moreau JJ (1988) Unilateral contact and dry friction in finite freedom dynamics. In: Moreau JJ, Panagiotopoulos PD (eds) Non-smooth mechanics and applications. CISM courses and lectures, vol. 302. Springer, Wien, pp 1–82
107. Jean M (1999) The non-smooth contact dynamics method. Comput Meth Appl Mech Eng 177:235–257
108. Stewart DE (2000) Rigid-body dynamics with friction and impact. Siam Rev C Soc Ind Appl Math 42(1):3–39
109. Génot F, Brogliato B (1999) New results on Painlevé's paradoxes. Eur J Mech A Solids 18:653–677
110. Payr M, Glocker C (2005) Oblique frictional impact of a bar: analysis and comparison of different impact laws. Nonlinear Dyn 41:361–383
111. Zhao Z, Liu C, Chen B (2008) The Painlevé paradox studied at a 3d slender rod. Multibody Sys Dyn 19:323–343
112. Liu C, Zhao Z, Chen B (2007) The bouncing motion appearing in a robotic system with unilateral constraint. Nonlinear Dyn 49:217–232
113. Lancioni G, Lencia S, Galvanetto U (2009) Non-linear dynamics of a mechanical system with a frictional unilateral constraint. Int J Non Lin Mech 44:658–674
114. Hunt KH, Crossley FRE (1975) Coefficient of restitution interpreted as damping in vibroimpact. ASME J Appl Mech 42(2):440–445
115. Shin K, Brennan MJ, Oh J-E, Harris CJ (2002) Analysis of disc brake noise using a two-degree-of-freedom model. J Sound Vib 254(5):837–848
116. Sinou J-J, Jézéquel L (2007) Mode coupling instability in friction-induced vibrations and its dependency on system parameters including damping. Eur J Mech Solid 26:106–122
117. Sinou J-J, Jézéquel L (2007) The influence of damping on the limit cycles for a self-exciting mechanism. J Sound Vib 304:875–893
118. Vahid O, Eslaminasab N, Golnaraghi MF (2009) Friction-induced vibration in lead screw systems: mathematical modeling and experimental studies. J Vib Acoust 131:021003
119. Erkorkmaz K, Kamalzadeh A (2006) High bandwidth control of ball screw drives. Ann CIRP 55(1):393–398
120. Kamalzadeh A, Erkorkmaz K (2007) Accurate tracking controller design for high-speed drives. Int J Mach Tools Manufact 47:1393–1400

Index

GPSR Compliance

The European Union's (EU) General Product Safety Regulation (GPSR) is a set of rules that requires consumer products to be safe and our obligations to ensure this.

If you have any concerns about our products, you can contact us on ProductSafety@springernature.com

In case Publisher is established outside the EU, the EU authorized representative is:

Springer Nature Customer Service Center GmbH
Europaplatz 3
69115 Heidelberg, Germany

Batch number: 10755292

Printed by Printforce, the Netherlands